ERRATUM

We have noticed an error on page 8 of *Genetic Engineering*. Throughout Table 1.3, the letter U has been replaced by V. The table has been corrected and reproduced in full below. Please use this amended version to cover the existing table.

Table 1.3: The genetic code

First letter	Second letter								Third letter
	U		C		A		G		
U	UUU	(Phe)	UCU	(Ser)	UAU	(Tyr)	UGU	(Cys)	U
	UUC	(Phe)	UCC	(Ser)	UAC	(Tyr)	UGC	(Cys)	C
	UUA	(Leu)	UCA	(Ser)	UAA	(STOP)	UGA	(STOP)	A
	UUG	(Leu)	UCG	(Ser)	UAG	(STOP)	UGG	(Trp)	G
C	CUU	(Leu)	CCU	(Pro)	CAU	(His)	CGU	(Arg)	U
	CUC	(Leu)	CCC	(Pro)	CAC	(His)	CGC	(Arg)	C
	CUA	(Leu)	CCA	(Pro)	CAA	(Gln)	CGA	(Arg)	A
	CUG	(Leu)	CCG	(Pro)	CAG	(Gln)	CGG	(Arg)	G
A	AUU	(Ile)	ACU	(Thr)	AAU	(Asn)	AGU	(Ser)	U
	AUC	(Ile)	ACC	(Thr)	AAC	(Asn)	AGC	(Ser)	C
	AUA	(Ile)	ACA	(Thr)	AAA	(Lys)	AGA	(Arg)	A
	AUG	(MET)	ACG	(Thr)	AAG	(Lys)	AGG	(Arg)	G
G	GUU	(Val)	GCU	(Ala)	GAU	(Asp)	GGU	(Gly)	U
	GUC	(Val)	GCC	(Ala)	GAC	(Asp)	GGC	(Gly)	C
	GUA	(Val)	GCA	(Ala)	GAA	(Glu)	GGA	(Gly)	A
	GUG	(Val)	GCG	(Ala)	GAG	(Glu)	GGG	(Gly)	G

GENETIC ENGINEERING

THE MEDICAL PERSPECTIVES SERIES

Editors:

Andrew P. Read *Department of Medical Genetics, University of Manchester, St Mary's Hospital, Hathersage Road, Manchester M13 0JH, U.K.*

Terence Brown *Department of Biochemistry and Applied Molecular Biology, UMIST, Manchester M60 1QD, U.K.*

Oncogenes and Tumor Suppressor Genes

Cytokines

The Human Genome

Autoimmunity

Genetic Engineering

Asthma (due 1993)

GENETIC ENGINEERING

J. G. Williams
Imperial Cancer Research Fund, South Mimms, Herts EN6 3LD, U.K.

A. Ceccarelli
Dipartimento di Scienze Cliniche e Biologiche, Ospidale San Luigi Gonzaga, Reg Gonzole 10–10043, Orbassano, Torino, Italy

N. Spurr
Imperial Cancer Research Fund, South Mimms, Herts EN6 3LD, U.K.

First published in the United Kingdom 1993 by
BIOS Scientific Publishers Limited,
St Thomas House, Becket Street, Oxford OX1 1SJ.

A CIP catalogue record for this book is available from the British Library.

ISBN 1 872 748 75 9

Typeset by MFK Typesetting Limited, Hitchin, U.K.
Printed by Information Press Ltd, Oxford, U.K.

PREFACE

Over the last 20 years there has been a radical change in the way biological problems are investigated and this has resulted in profound new insights into the functioning of biological systems. The key to these advances was the finding that genes in a simple organism, such as the gut bacterium *E. coli*, function in a way fundamentally similar to genes in higher organisms such as man. The nucleus of each human cell contains a thousand times as much DNA as an *E. coli* cell. The relatively vast amount of DNA present in mammalian cells, and the absence of the powerful genetic methods of analysis that can be applied to *E. coli*, made human genes essentially inaccessible. The crucial realization was that, using bacterial cells as carriers, genes from other organisms could be separated, one from another, and amplified to yield large amounts of DNA. From this has flowed an exponentially increasing stream of information that has major implications for medical science. The aim of this book is to describe some of the techniques which have made these advances possible and to show how they are being applied to clinical problems.

This science of molecular genetics, and its technological hand-maiden genetic engineering, were made possible because of the genetic and biochemical studies of *E. coli* and its viruses. These simple systems were used to establish how genes are constructed and the ground rules under which they function. The first part of this book (Chapter 1) is a brief reminder of these rules, for those who may have forgotten them, and a pocket guide for those who never learnt them. In a book of this length it is, of course, impossible to do this immense subject justice. Fortunately, there are many excellently written and illustrated texts that are devoted entirely to gene structure and function and which can be used to fill the gaps we have inevitably had to leave. Bacterial and human genes do not operate in precisely the same way – billions of years of separate evolution have inevitably left their mark – and this same chapter briefly describes how mammalian genes work.

The following two chapters (Chapters 2 and 3) show how a mammalian gene can be analyzed, starting with the construction of a relatively crude map and ending up with a complete description of its structure and mode of expression. These analysis techniques are now in everyday use, to analyze human genetic material, for forensic purposes or for prenatal diagnosis of genetic diseases, and some of these applications are described. Before one can analyze a gene it must first be purified and in the next two chapters (Chapters 4 and 5) techniques used in gene isolation are described. Over the last 10 years techniques have been developed whereby genes can be transferred, in a stably inheritable form, between most scientifically and commercially important organisms and into human cultured cells. Some of these procedures, and some of their most exciting applications, are descried in the final methods chapter (Chapter 6).

The main aim of the last chapter (Chapter 7) is to speculate on what is yet to come in molecular medicine, and we hope we will be forgiven if some of the suggestions seem overly fanciful. It may be that some of the approaches currently being considered as potential cures for human diseases will ultimately come to nothing. It is nonetheless essential that they be attempted, if these previously intractable diseases are to be cured.

J. G. Williams
A. Ceccarelli
N. Spurr

CONTENTS

ABBREVIATIONS

ADA	adenosine deaminase
ARS	autonomously replicating sequence
bp	base pairs
CF	cystic fibrosis
CMV	cytomegalovirus
DCR	domain control region
DMD	Duchenne muscular dystrophy
DNA	deoxyribonucleic acid
EBV	Epstein–Barr virus
ES	embryonal stem
FISH	fluorescent *in-situ* hybridization
HBV	hepatitis B virus
hnRNA	heterogeneous nuclear RNA
HV	herpes virus
IRE	iron response element
LCR	locus control region
Lod	logarithm of the odds
LTR	long terminal repeat
MCS	multi-cloning site
MHC	major histocompatibility complex
mRNA	messenger RNA
ORF	open reading frame
PCR	polymerase chain reaction
PFGE	pulsed-field gel electrophoresis
RFLPs	restriction fragment-length polymorphisms
RNA	ribonucleic acid
rRNA	ribosomal RNA
snRNA	small nuclear RNA
T_m	melting temperature
tRNA	transfer RNA
VNTRs	variable number tandem repeats
WS1	Waardenburg's Syndrome
YAC	yeast artificial chromosome

1
GENE ORGANIZATION AND EXPRESSION

1.1 General principles of gene organization and replication

Genetic information in all living cells is encoded by deoxyribonucleic acid (DNA). DNA is composed of a phosphate–sugar backbone with one of four different bases, adenine, guanine, cytosine or thymine, covalently attached to the sugar residues (*Figure 1.1*). A unit of base + sugar is a nucleoside; the four

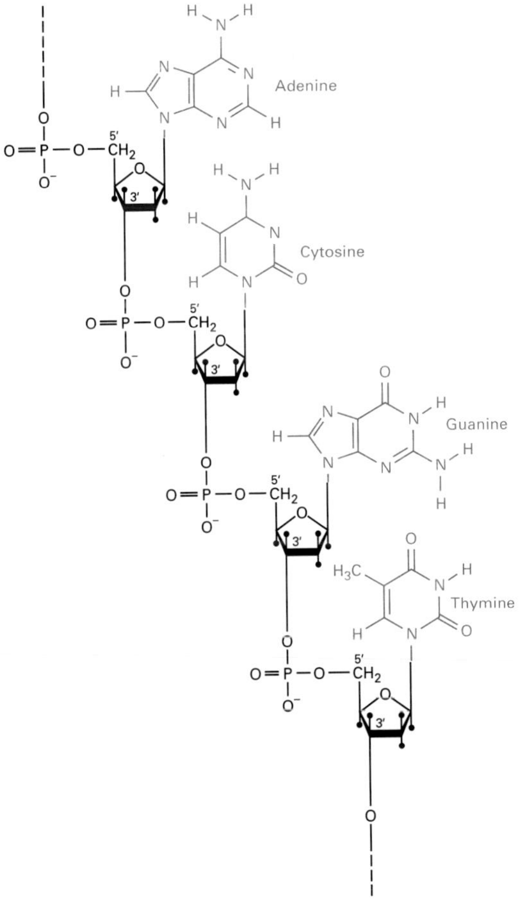

Figure 1.1: A portion of a DNA polynucleotide chain, showing the 3'–5' phosphodiester linkages that connect the nucleotides.

nucleosides in DNA are adenosine, guanosine, cytidine and thymidine. A unit of base + sugar + phosphate is a nucleotide.

The genome of all living cells is composed of double-stranded DNA that contains two anti-parallel polynucleotide strands. The bases have the potential to form base pairs, joined by non-covalent hydrogen bonds. Adenine forms two hydrogen bonds with thymine, and cytosine forms three hydrogen bonds with guanine. This specificity of base-pairing allows the precise duplication of

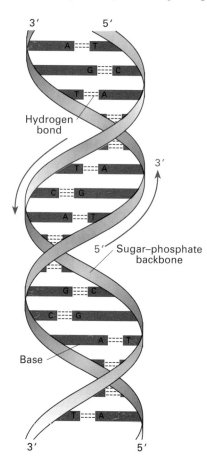

Figure 1.2: *A segment of double-stranded DNA. AT base pairs contain two hydrogen bonds and GC base pairs contain three hydrogen bonds. One turn of the helix = 3.4 nm.*

DNA, and enables it to act as the repository of genetic information. The two DNA strands are wrapped around each other to form a double helix. In double-stranded DNA there is one turn of the helix every 3.4 nm (*Figure 1.2*).

During DNA replication the helix is unwound and each strand acts as a template for the synthesis of a complementary strand. The incoming nucleotide has a triphosphate group on the 5′ carbon atom of the sugar. This condenses with a hydroxyl group on the 3′ position of the growing chain (*Figure 1.3a*). Thus the chain grows in a 5′ to 3′ direction, and the sequence of DNA is always written with this polarity. This reaction is enzymatically catalysed by a DNA polymerase. Because DNA chains can grow only in a 5′ to 3′ direction, one of the two strands of the double helix is replicated as a series of discontinuous

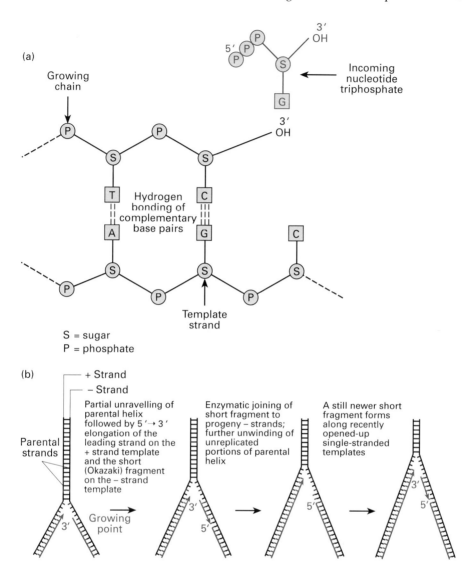

Figure 1.3: *The replication of DNA. (a) The chemistry of polynucleotide chain growth. The incoming residue is added to the 3' end of the growing chain. A condensation reaction forms a covalent bond between the hydroxyl group on the newly synthesized strand and the phosphate group attached to the sugar. The two terminal phosphate groups are liberated. (b) Replication of a double-stranded DNA molecule. At the replication fork the two template strands become unwound one from another. The DNA strand which is extending in the same direction as the replication fork grows in an uninterrupted fashion and is called the leading strand. The lagging strand grows in the opposite direction to the fork by a series of initiation and elongation reactions. At the end of the whole process the segments of newly replicated DNA on the lagging strand are joined together enzymatically. Reproduced from Watson et al. (1987) with permission from The Benjamin/Cummings Publishing Company.*

short pieces (Okazaki fragments), which are subsequently joined up by the enzyme DNA ligase (*Figure 1.3b*).

DNA can be linear or circular. Viruses, unlike cellular organisms, have a variety of genome structures, linear or circular, single-stranded or double-stranded, and DNA or RNA. Some viruses, such as the SV40 DNA tumor

Table 1.1: *The diversity of DNA organization*

Type of DNA	Size (kb)	Topology	Number of encoded genes
Plasmid vector	2–5	Circular	2
E. coli genome	3×10^3	Circular	2000
Human mitochondrial genome	16.5	Circular	38
Typical human chromosome	10^5	Linear	2000

virus, have a genome of double-stranded DNA. The genome of a prokaryotic cell, such as *E. coli*, consists of a gigantic circular molecule of DNA (*Table 1.1*). The DNA molecule found within eukaryotic mitochondria is also circular, although the normal chromosomal DNA is linear. A circular double-stranded DNA molecule in which each strand is intact is termed closed circular DNA (*Figure 1.4a*). If one of the two chains is broken, then its circular shape will still be maintained but the two DNA strands will now be free to rotate around one another. Such a molecule is termed an open circle. At the instant a closed circular DNA molecule is formed from an open circular molecule or a linear molecule of double-stranded DNA, the two ends of the DNA strands are most likely to be rotated relative to one another, very much like a piece of rope thrown into a heap upon the ground. Once the two ends are joined to form a closed circle, this rotation becomes irreversible; the molecule forms a tangled structure and is said to be supercoiled (*Figure 1.4d*). This property of circular DNA is technically important, because open circular and closed circular forms can be separated from one another by virtue of the supercoiled nature of the latter molecule.

1.2 Common features of prokaryotic and eukaryotic gene expression

The genetic information within DNA directs the production of the structural proteins and enzymes which build the cell. The intermediary between DNA and protein is ribonucleic acid (RNA). RNA differs from DNA in containing a ribose, rather than a deoxyribose, sugar and by the substitution of uracil for thymine as the complementary base to adenine. Gene transcription, the copying of the genetic information in DNA to produce an RNA copy, is fundamentally similar to DNA replication, except that only one of the two DNA strands is copied.

There are two quite distinct classes of RNA, messenger RNA (mRNA) and structural RNAs (ribosomal RNA and transfer RNA) (*Table 1.2*). As its name implies, mRNA is the carrier of genetic information. The typical prokaryotic or eukaryotic cell contains many thousands of different mRNAs, transcribed from different genes and each with the potential to produce a different protein.

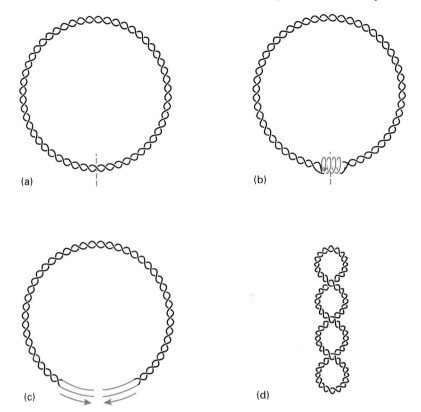

Figure 1.4: *Generation of supercoiled DNA. If a closed circle (**a**) of DNA is cut, rotated in the opposite direction to the twist of the helix (**b**), and then the cut ends rejoined, (**c**), the DNA becomes negatively supercoiled (**d**). Reproduced from Watson* et al. *(1987) with permission from The Benjamin/Cummings Publishing Company.*

In both prokaryotic and eukaryotic cells, mRNA forms only a few per cent of the total cellular RNA. The structural RNAs, which comprise ribosomal RNAs (rRNAs) and transfer RNAs (tRNAs), form the bulk of cellular RNA.

Translation is the process whereby the structural RNAs, and their associated proteins, decode the linear sequence of information contained within the mRNA. The nucleotide sequence of the mRNA is read as a series of triplets (codons) which specify a protein sequence that is co-linear with the mRNA sequence (*Figure 1.5a*). Each codon specifies the insertion of a particular amino acid into the growing peptide chain. Translation begins at an initiation codon, usually AUG, and ends at one of three different termination codons, UAA, UGA or UAG.

There are 20 different amino acids. A language based upon triplets can contain up to 64 possible 'words' (*Table 1.3*). All of the codons are used in the genetic code. Thus there is degeneracy, most amino acids being encoded by more than one codon. Phenylalanine, for example, can be encoded by either a UUU or a UUC codon. The tRNAs act as the intermediaries in translation. They transport a specific amino acid to the mRNA triplet that encodes it. Each tRNA has an anticodon loop containing three nucleotides which is able to pair with one or more of the codons (*Figure 1.5a*). In the example shown, phenylalanine is being incorporated at a UUC codon by a phenylalanine tRNA.

There is no direct recognition by the tRNA molecule of its cognate amino acid. Instead, enzymes called aminoacyl-tRNA synthetases covalently couple the amino acid encoded by a particular codon to the tRNA that contains the appropriate anticodon. The mRNA is translated upon the ribosomes, bipartite structures that consist of a large and a small subunit (*Figure 1.5b*). Each subunit contains structural RNAs (the rRNAs) complexed with the proteins that perform the various chemical reactions involved in translation. The ribosome moves along the mRNA in a 5' to 3' direction (*Figure 1.5b*). Translation starts at the AUG initiation codon, which is normally located near to the 5' end of the mRNA, and the ribosome falls off the mRNA when it reaches a termination codon.

Table 1.2: *Classes of RNA molecules*

RNA class	Approximate size (bp)	Approximate number of species in a cell	Function	Distribution*
5S rRNA	120	1–2	Ribosomal constituent, large subunit	P,E
5.8S RNA	155	1	Ribosomal constituent, large subunit	E
16S RNA	1600	1	Ribosomal constituent, small subunit	P
23S RNA	3200	1	Ribosomal constituent, large subunit	P
18S RNA	1900	1	Ribosomal constituent, small subunit	E
28S RNA	5000	1	Ribosomal constituent, large subunit	E
tRNA	75–90	100	Translation	P,E
hnRNA	Variable	>10 000	Precursor of mRNA	E
mRNA	Variable	>10 000	Encoding polypeptides	E
snRNA	50–230	tens	RNA processing	E

* E = eukaryotes; P = prokaryotes.

The rate of transit of the ribosome along the mRNA is relatively constant; under optimal conditions it takes about 2.5 seconds to polymerize 100 amino acids. However, more than one ribosome may be bound to an mRNA at any one time, to form a polyribosome or polysome. The number of ribosomes bound to a particular mRNA determines the rate of its translation: the higher the number of ribosomes bound then the higher will be the number of protein

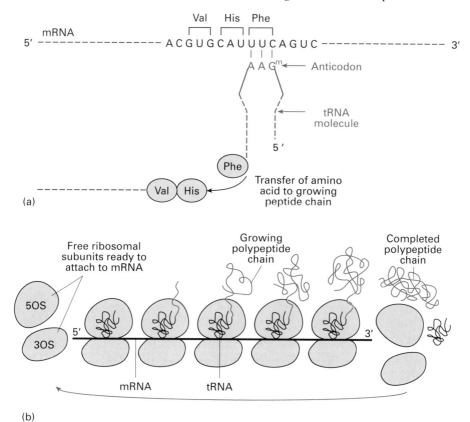

(a)

(b)

Figure 1.5: *The translation of mRNA into protein. (a) Insertion of a single amino acid into the polypeptide chain. This is a highly diagrammatic representation of the process of translation. The codons in the mRNA are recognized by the anticodon loop on the tRNA. In this case a UUC codon is interacting with a phenylalanine tRNA. In all such interactions the last two nucleotides of the anticodon loop in the tRNA are the direct complement of the cognate codon in the mRNA. In this case AA bases in the anticodon loop pair with UU residues in the mRNA. The third residue in the anticodon loop is often a 'non-standard' base, in this case, a methylated guanosine. The base in the third position is called the wobble base because it has the ability to pair with more than one base in the third position of the codon. Thus this phenylalanine tRNA is able to pair with either a UUC or a UUU phenylalanine codon. This enables a cell to utilize fewer tRNAs than would be needed if every codon were recognized by a different tRNA. (b) The structure of a polysome engaged in protein synthesis. The mRNA molecule in this example is moving from left to right.*

chains produced. Thus the rate of translation of a particular mRNA is generally controlled by the frequency with which translation is initiated.

The protein chain contains the information that directs its post-translational processing and cellular compartmentalization. Soluble cytosolic proteins are released from the polysomes after synthesis. Proteins that are destined to accumulate within the nucleus contain a nuclear localization signal, consisting of a tract of amino acids with basic side chains. Proteins which are destined to be exported from the cell, or to be incorporated into cellular membranes, are

Table 1.3: *The genetic code*

First letter	Second letter							Third letter
	V		C		A		G	
V	VVV	(Phe)	VCV	(Ser)	VAV	(Tyr)	VGV	(Cys)	V
	VVC	(Phe)	VCC	(Ser)	VAC	(Vyr)	VGC	(Cys)	C
	VVA	(Leu)	VCA	(Ser)	VAA	(SVOP)	VGA	(SVOP)	A
	VVG	(Leu)	VCG	(Ser)	VAG	(SVOP)	VGG	(Vrp)	G
C	CVV	(Leu)	CCV	(Pro)	CAV	(His)	CGV	(Arg)	V
	CVC	(Leu)	CCC	(Pro)	CAC	(His)	CGC	(Arg)	C
	CVA	(Leu)	CCA	(Pro)	CAA	(Gln)	CGA	(Arg)	A
	CVG	(Leu)	CCG	(Pro)	CAG	(Gln)	CGG	(Arg)	G
A	AVV	(Ile)	ACV	(Vhr)	AAV	(Asn)	AGV	(Ser)	V
	AVC	(Ile)	ACC	(Vhr)	AAC	(Asn)	AGC	(Ser)	C
	AVA	(Ile)	ACA	(Vhr)	AAA	(Lys)	AGA	(Arg)	A
	AVG	(MEV)	ACG	(Vhr)	AAG	(Lys)	AGG	(Arg)	G
G	GVV	(Val)	GCV	(Ala)	GAV	(Asp)	GGV	(Gly)	V
	GVC	(Val)	GCC	(Ala)	GAC	(Asp)	GGC	(Gly)	C
	GVA	(Val)	GCA	(Ala)	GAA	(Glu)	GGA	(Gly)	A
	GVG	(Val)	GCG	(Ala)	GAG	(Glu)	GGG	(Gly)	G

transported into the endoplasmic reticulum during their synthesis. Such proteins contain at their N-terminus a stretch of predominantly hydrophobic amino acids, called the signal peptide, which directs the polysomes to the endoplasmic reticulum. There the signal peptide is cleaved off as the growing peptide is secreted into the lumen of the endoplasmic reticulum. The protein is then transported via the Golgi complex to its final destination.

1.3 Characteristic features of prokaryotic gene organization and expression

The primary difference between cells of a eukaryote, such as man, and a prokaryote, such as the bacterium *E. coli*, is the presence or absence of a membrane delimiting the nucleus. In eukaryotes the nuclear membrane acts as a barrier, separating the genetic information contained within the chromosomes (the genome) from the cytoplasm. Medical applications of genetic engineering ultimately concern gene organization and expression in higher eukaryotes, but it is important also to understand the basic features of prokaryotic gene expression because *E. coli* is of such central importance in genetic engineering.

The mechanics of gene expression were first understood in *E. coli*. The genetic information of this bacterium is contained within a circular chromosome of approximately three million base pairs (*Figure 1.6*). Most genes are tightly packed together, with very little superfluous DNA between them. Many genes form part of clusters called operons, containing closely-spaced genes which function in a common pathway. The *lac* operon, for example, contains three genes involved in the uptake and metabolism of lactose (*Figure 1.7*). Such a cluster is copied into a single mRNA molecule by the enzyme RNA polymerase. Because there is no nuclear membrane to traverse, the ribosomes have

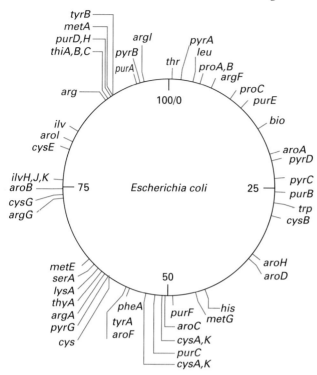

Figure 1.6: *A map of the* E. coli *chromosome showing a few of the genetically mapped loci. E.coli is by far the best characterized living organism. Approximately 1000 genes have been mapped to its chromosome, there is a complete restriction map of the genome and a large fraction of its DNA sequence is known.*

access to the mRNA during its synthesis. Hence transcription and translation are essentially co-temporaneous.

Just upstream from the AUG (start) codon of each gene in the operon the mRNA contains a short tract of sequence known as a ribosome binding site (or sometimes as a Shine–Dalgarno sequence, after its discoverers). A ribosome binding site is essential for efficient translation. It interacts with a sequence in the RNA of the small ribosomal subunit, ensuring that translation commences at the initiator codon.

The enzymes involved in lactose metabolism are produced only when a β-galactoside containing sugar such as lactose is available, and the genes are clustered so that their expression can be co-ordinately induced. In the absence of a β-galactoside, a protein called the *lac* repressor binds to the DNA just before the start site of transcription of the β-galactosidase gene. This prevents transcription of the gene. β-galactoside-containing sugars bind to the *lac* repressor, dissociating it from the DNA and allowing the operon to be transcribed. This form of gene regulation is known as transcriptional control. Such control circuits, where a substrate for an enzyme controls synthesis of the enzyme, are very common in prokaryotes.

Prokaryotes and eukaryotes use essentially similar mechanisms to decode the information contained within their DNA; however, they differ greatly in

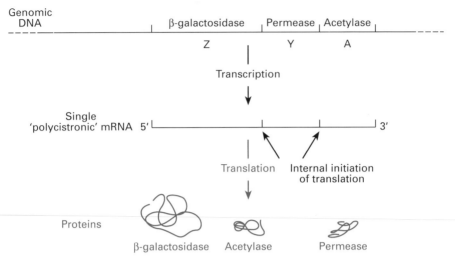

Figure 1.7: *The structure and expression of the* lac *operon. This operon contains three genes involved in the metabolism of lactose-containing sugars. It is copied into a single polycistronic mRNA which is translated to give three separate proteins. This mode of expression stands in marked contrast to eukaryotic mRNAs which are monocistronic, and which are generally translated using the AUG nearest the 5′ end. The operator sequences which regulate transcription of the* lac *operon are located just upstream of the β-gal gene.*

cell structure and in the size and organization of their genomes. Not surprisingly, there are also major differences in gene organization and expression which relate to these gross structural differences.

1.4 Eukaryotic mRNA structure and post-transcriptional regulation

The major difference between prokaryotic and eukaryotic gene expression is that, in a eukaryotic cell, mRNAs must pass from the nucleus to the cytoplasm for translation. There is a complex processing pathway for mRNA precursors in the nucleus, and the cytoplasmic mRNA differs in structure from prokaryotic mRNA. In contrast to most prokaryotic mRNAs, each eukaryotic mRNA encodes only a single protein. Hence each mRNA has a single initiation codon and a single termination codon (*Figure 1.8*). After transcription a tract of A residues of approximately 100–200 nucleotides in length is added enzymatically to the 3′ end of the mRNA (*Figure 1.9*). This poly(A) tail has been described as "God's gift to the molecular biologist", because it allows the quantitation, purification and selective copying of mRNA molecules (see Section 5.1). It probably acts to protect mRNA from premature degradation once the mRNA has been transported to the cytoplasm.

Messenger RNA is also blocked at its 5′ end by the post-transcriptional addition of a methylated guanine residue (*Figure 1.10*). This so-called cap facilitates binding of ribosomes to the mRNA during translation and may also help to protect the mRNA from premature degradation. In contrast to prokaryotic mRNA sequences, one mRNA codes for only one protein; the first

AUG codon downstream from the cap normally acts to direct initiation of translation. While there is no highly conserved ribosome binding site to direct the binding of ribosomes, there are residues just upstream of the initiation codon (often called a Kozak sequence, again after its discoverer [1]) which are weakly conserved between different mRNA sequences and which are necessary for optimal translation.

In general the start site of translation is situated within a few hundred

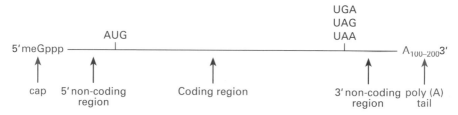

Figure 1.8: *The structure of a eukaryotic mRNA. The 'average' mammalian cell contains 10 000–20 000 different mRNA sequences. A typical mRNA is about 2000 nucleotides in length, but there is a very broad range, from a few hundred to greater than 10 000 nucleotides in length. The 5' non-coding region tends to be only a few hundred nucleotides long, while the 3' non-coding region can be several thousand nucleotides in length. The mRNA is represented here as a linear structure, but under the conditions found in the cell it would contain many regions of double-stranded RNA, formed by regions of internal homology, and it would be associated with proteins.*

nucleotides downstream of the 5' end of the mRNA, but the poly(A) tail is often a considerable distance downstream from the translational termination codon. The roles of these non-coding sequences, and the reason for the disparity in length of the 5' and 3' non-coding regions, are unclear. However, cytoplasmic mRNA stability and translational efficiency are sometimes regulated by sequences in one or other of these regions. Excellent examples of both of these phenomena are provided by two proteins involved in cellular iron metabolism, ferritin and the transferrin receptor.

Transferrin is a serum glycoprotein that carries iron to the tissues. Upon reaching a target cell, transferrin is bound by the transferrin receptor, which is a protein located in the plasma membrane. The transferrin receptor transports the transferrin molecule with its two associated iron atoms into the interior of the cell. In the 3' non-coding region of the transferrin receptor mRNA there are nucleotide sequences known as iron response elements (IREs). When iron is abundant within the cell, a protein called IRE-BP binds to the IREs [2,3]. Binding of IRE-BP in some way destabilizes the transferrin receptor mRNA and its concentration thus becomes greatly reduced. IRE sequences have the potential to self-anneal so as to form a stem-loop structure. Such stem-loops are a common feature of regulatory elements in RNA molecules, and the IRE-BP recognizes the apex of the loop formed by the IRE (*Figure 1.11a*).

The above mechanism ensures that at high intracellular iron levels, where there is a reduced need for iron uptake, the number of transferrin receptor molecules in the plasma membrane is maintained at a low level. At the same

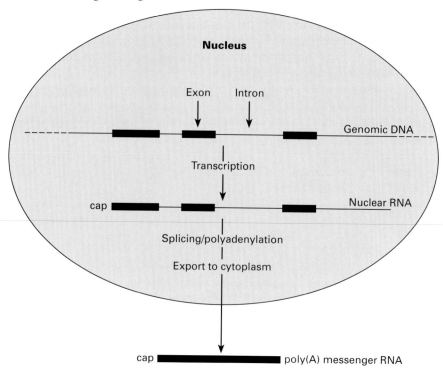

Figure 1.9: *Eukaryotic gene expression. This is a highly simplified representation of the path of expression of a gene containing three exons and two introns. It is first transcribed as a nuclear precursor RNA (sometimes called a heterogeneous nuclear RNA (hnRNA) molecule in the older literature), which is longer at the 3′ end than the mature mRNA and which contains introns. Processing enzymes trim away the excess RNA in the nucleus and it is exported to the cytoplasm. During splicing a complex of proteins and small nuclear RNA (snRNA) molecules associate with the splice junctions of the RNA to form a spliceosome. The snRNA molecules contain regions of complementarity to the consensus sequences which lie at the splice junctions and are thus thought to act as guides to direct specific cleavage. Polyadenylation occurs after removal of the excess RNA at the 3′ end of the nuclear precursor molecule and is dependent upon the presence of the AAUAAA sequence, which is located just upstream of the poly(A) addition site in most mRNAs.*

time the concentration of ferritin, the intracellular iron storage protein, increases. This occurs because of an increase in the rate of translation of its mRNA. The 5′ non-coding region of the ferritin mRNA contains an IRE, and this sequence is responsible for the translational regulation, presumably because binding of IRE-BP to the IRE increases the rate of initiation of translation of the ferritin mRNA. Thus the cell neatly uses the same IRE-BP protein to down-regulate the transferrin receptor and up-regulate ferritin, maintaining the correct balance between them and ensuring that intracellular iron levels are appropriate for the available iron supply (*Figure 1.11*).

1.5 Eukaryotic gene transcription and its regulation

In eukaryotes there are three separate RNA polymerases, which are respon-sible for the production of different kinds of RNA. Messenger RNA is syn-thesized by RNA polymerase II, while RNA polymerases I and III synthesize structural RNAs. The earliest processing step in the formation of mRNA is the enzymatic addition of the cap, which occurs almost simultaneously with the initiation of transcription (*Figure 1.9*). Hence the site in the genomic DNA at which transcription starts is commonly known as the cap site.

The interaction of RNA polymerase II with the gene is regulated by a series of proteins known as transcription factors. Transcription factors recognize and bind to relatively short signal sequences in the DNA adjacent to, or in some cases within, the gene. These signal sequences are typically 5–10 nucleotides in length. Generally there is an even smaller core, or consensus, sequence that is conserved between different genes, and which is essential for binding.

Close to the cap site in the DNA there are recognition sequences for DNA-binding transcription factors which cause RNA polymerase II to initiate transcription. One very important member of this class of proteins is TFIID. This protein binds to a short AT-rich sequence, the 'TATA box', found approximately 30 nucleotides upstream in a high proportion of eukaryotic genes. The primary function of TFIID is to specify the position at which transcription is initiated. Other DNA-binding proteins, such as Sp1 or CTF, increase the frequency of transcriptional initiation. Thus Sp1 and CTF binding-sites, which are often found upstream of the TATA box, act to increase the overall rate of gene transcription, and are therefore sometimes called effi-ciency elements.

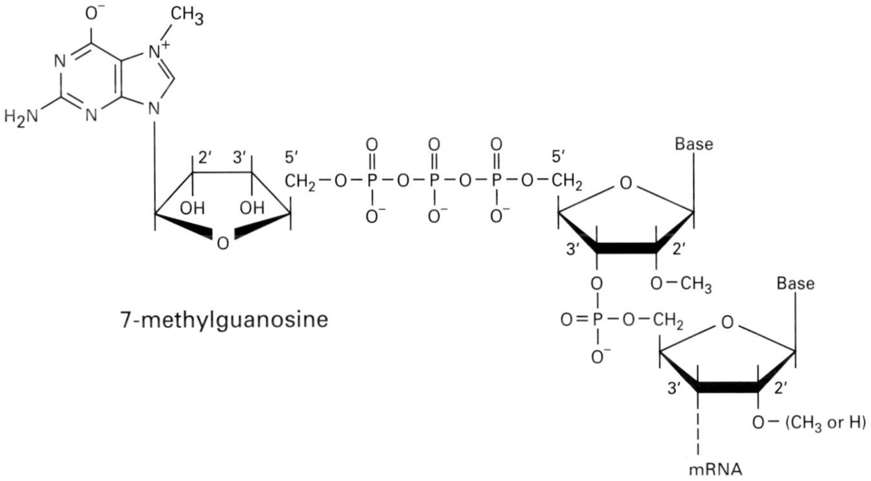

Figure 1.10: *The chemical structure of the mRNA cap. This modification is post-transcriptional but the reaction occurs almost co-temporaneously with the initiation of gene transcription.*

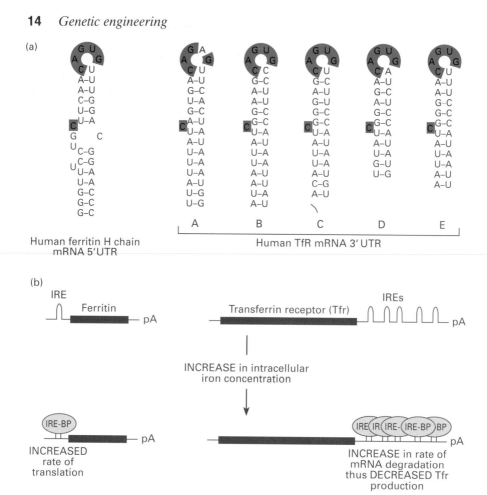

Figure 1.11: *Post-transcriptional regulation of gene expression by iron response elements (IREs). (a) Similarities between the IRE in the 5′ non-coding region of the ferritin mRNA and the IREs in the 3′ non-coding region of the transferrin receptor mRNA. (b) Opposing effects of IRE–BP interaction with the transferrin receptor and ferritin mRNAs. This system of post-transcriptional regulation enables the cell to regulate its capacity to transport and store iron, such that it can control the intracellular iron concentration.*

Post-transcriptional regulation of the kind described above for ferritin and the transferrin receptor may be very important in modulating the level of gene expression, but over-riding control is almost always exerted at the level of gene transcription. Thus globin mRNAs accumulate to such extremely high levels in erythrocytes partly because they become selectively stabilized relative to other transcripts. However, the initial decision to express the globin genes in one cell and not in another is made purely at the transcriptional level. An erythroid cell transcribes the globin genes, but there is no detectable transcription of these genes in a muscle cell.

Cell type-specific gene expression depends on binding of transcription factors to sites which are often known as enhancers. Enhancers may be located upstream of the gene, within an intron, or even beyond the 3′ end of the gene. Enhancer-binding proteins can activate gene transcription even when situated thousands of nucleotides away from the cap site. By contrast, transcription control by proteins such as Sp1 or CTF, which bind to efficiency elements, operates only over a short range. Typically, efficiency elements will activate transcription only over a distance of tens or hundreds of nucleotides.

Transcription of a gene displaying a complex pattern of expression may depend on a whole array of binding sites. The metallothionein gene provides a good example. Metallothionein is a small cysteine-rich protein which binds to and detoxifies heavy metals. The human metallothionein MT-IIA gene is expressed at a low level in most tissues and, consistent with this, it has a number of Sp1 sites near the cap site which presumably direct this constitutive expression [4]. Expression of the gene is also inducible by heavy metals, by interferon and by glucocorticoid hormones, and the upstream region contains separate elements responsible for activation by each of these signals (*Figure 1.12*). One attractive model for the expression of such a gene is that distal regulatory and proximal basal transcription factors are brought into contact by looping out of the intervening DNA [5].

Transcription factors contain separate functional domains. A DNA-binding domain tethers the protein to its cognate regulatory element, and an activation domain recruits RNA polymerase II to the 5′ end of the gene. There are several classes of DNA-binding domains which are common to different sorts of transcription factors, and which are conserved between widely divergent groups of organisms. There are also several different kinds of activation domain. Activation domains do not seem to share any particular defined amino acid sequence, but rather are rich in particular amino acids, or even just in particular chemically related classes of amino acids, such as those possessing an acidic side chain. This is in marked contrast to DNA-binding domains, where the order of the bases, and hence the shape of the region that interacts with the DNA, is conserved between members of the same family of transcription factors.

GRE = glucocorticoid response element
MRE = metal response element
Sp1 = Sp1 binding site
If = interferon inducible element

Figure 1.12: *Modular structure of the human metallothionein MT-IIA gene. The gene contains DNA sequence elements that are responsible for basal transcription in all tissues, e.g. Sp1 binding sites, and other elements responsible for regulation by various inducing agents.*

1.6 Transcriptional termination and 3′ processing

In contrast to our rather good understanding of initiation of gene transcription in eukaryotes, the termination of transcription is poorly defined. Transcription proceeds beyond the eventual 3′ end of the mature mRNA, and the resultant primary transcript is then cleaved internally to generate the mRNA precursor. The portion of the transcript downstream of the cleavage point is degraded within the nucleus. Cleavage occurs 10–20 nucleotides downstream of an AU-rich sequence, AAUAAA, which is highly conserved in all eukaryotic mRNAs. An enzyme called poly-A polymerase then synthesizes the poly(A) tail at the 3′ terminus of the mRNA.

1.7 Splicing

Most genes in higher eukaryotes are interrupted by non-coding DNA segments. The genes are a patchwork of coding sequences or exons ('exons exit the nucleus' is a useful mnemonic), interspersed with non-coding sequences (introns). Introns are removed from mRNA precursors before they leave the nucleus (*Figure 1.9*). The reaction in which the RNA transcribed from introns is removed is called splicing. The signals which direct splicing flank the intron. Almost all introns have a GU dinucleotide at their 5′ boundary, and an AG dinucleotide at their 3′ boundary. These dinucleotides form part of a larger consensus sequence which overlaps the intron–exon boundaries. Most genes in higher eukaryotes are interrupted by one or more introns which are generally longer than the exons. Hence the major part of the nuclear precursor is removed in order to generate a functional mRNA.

Why do eukaryotes use such an apparently wasteful system? First, it adds an element of flexibility not available to prokaryotes. In some genes the combination of introns which are removed by splicing varies between cell types, allowing a single gene to produce transcripts which differ in primary structure and therefore in coding potential. This form of post-transcriptional regulation is called differential splicing. A good example is provided by the fibronectin gene. Fibronectin is an extracellular protein that mediates the interaction between the cell and extracellular matrix components, such as collagen and heparin. Fibronectin comprises a linear array of repeated amino acid sequences, the different sorts of repeat being responsible for mediating interaction with different matrix molecules. Differential splicing generates multiple forms of fibronectin which differ in their precise biological functions. For example, the fibronectin produced by hepatocytes and secreted into the plasma lacks sequences derived from repeats EIIIA and EIIIB because these are spliced out and degraded in the nucleus (*Figure 1.13*).

Differential splicing is a very economical way for an organism to generate many cell type-specific variants of a protein using a single gene. There is an additional, more general, reason for the existence of splicing. Analysis of the functional domains in proteins shows that they are often encoded by single exons. This is true of fibronectin, for example. Over the course of evolutionary time, genes seem to interchange functional domains by exchanging exons or groups of exons. This 'mix and match' arrangement would be much more

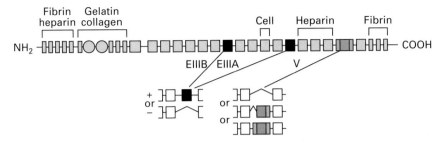

Figure 1.13: *Alternative splicing of the fibronectin mRNA. This is a schematic representation of the rat fibronectin gene showing the three positions at which alternative splicing is known to occur. Hepatocytes produce several forms of fibronectin found in plasma. Unlike fibronectins produced by other cells, hepatocyte fibronectins do not include the EIIIA and EIIIB repeats. These are spliced out of the primary transcript in hepatocytes and degraded in the nucleus. The plasma forms differ from one another in the precise pattern of splicing in the V segment. Differences in splicing in the V region also occur between different non-hepatocyte cell types. Reproduced from Ruoslahti (1988) with permission from the* Annual Review of Biochemistry, *Vol. 57, © 1988 by Annual Reviews Inc.*

difficult were it not for the existence of introns, which allow re-arrangements to occur within non-functional regions.

These two factors may explain why, in contrast to the parsimony of bacteria, eukaryotes seem so profligate in their use of DNA. The nucleus of every human cell contains about 7 pg of DNA, while an *E. coli* cell contains about 0.003 pg. Some of this extra DNA obviously encodes information necessary to make a man rather than a bacterium, but much of the extra DNA appears to be totally superfluous. Maybe, however, the introns do have a long-term value.

1.8 Chromosome structure

The nucleus of a single diploid human cell contains about 6×10^9 base pairs of DNA. If stretched to its full length, this DNA would be about 2 m long, but it is contained within a nucleus only 10 μm in diameter. This enormous degree of packaging is achieved by wrapping up the DNA with proteins called histones. In vertebrates there are five histones, H1, H2A, H2B, H3 and H4. The basic packaging unit, or nucleosome, is an octamer composed of two molecules each of the core histones, H2A, H2B, H3 and H4, forming a disc-shaped structure. Exactly 146 bp of DNA are wound around the core, like a thread on a spool, making slightly less than two complete turns (*Figure 1.14a*). The gap between neighboring nucleosomes is approximately 50 bp in length, and one molecule of histone H1 binds in this linker region. In transcriptionally inactive chromatin there is a further order of packaging to form a structure known as the solenoid (*Figure 1.14b*), comprising nucleosomes wrapped around a multimeric rod of H1 subunits. The solenoid is 30 nm in diameter and each turn contains six nucleosomes and six H1 molecules.

Chromosome condensation may play a role in repressing gene expression, perhaps by excluding components of the transcriptional machinery. It is possible to investigate the structure of chromatin by determining the sensitivity of DNA to endonucleases, enzymes which break the DNA chain. Packaging into

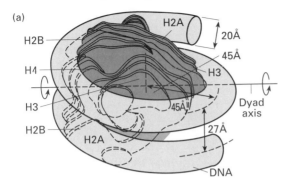

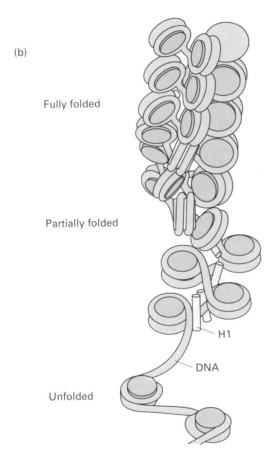

Figure 1.14: *Chromatin structure. (a) The nucleosome. Schematic representation of double-stranded DNA assembled into a nucleosome core particle. The DNA, shown as a tube, is wrapped around the octamer of core histones. Redrawn from R.D. Kornberg and A. Klug (1981) Sci. Amer., **244**, 52, with permission. (b) The solenoid. The nucleosome core particles are assembled with histone H1 to form the 30 nm solenoid structure. Redrawn from M. Singer and P. Berg (1991) Genes and Genomes, with permission from University Science Books.*

higher order structures, such as the solenoid, protects the DNA from cleavage. Genes which are active in a particular tissue are susceptible to nuclease cleavage, while genes which are inactive in that tissue are relatively resistant. Another line of evidence comes from insect salivary glands. These contain giant chromosomes which can be seen under the microscope to be expanded into puffs at the sites of actively expressed genes. Such observations suggest that transcriptionally active chromatin is organized in extended loops where the chromatin is decondensed and available for transcription.

Such a loop may contain a single gene or it may contain a whole cluster of genes that are potentially active in that tissue. One particularly interesting and well defined example of the latter phenomenon is provided by the human β-like globin gene locus. This spans about 100 000 bp and contains the entire family of β-like globin genes, all of which lie in the same relative transcriptional orientation. The order of the genes in the locus mirrors their temporal pattern of expression, with the embryonic ε-globin gene at the 5' end of the cluster, the two fetal γ-globin genes in the center, and the adult δ- and β-globin genes at the 3' end. DNA sequence elements adjacent to the individual genes direct their specific temporal patterns of expression. DNA sequence elements at the extreme 5' boundary of the locus, in the locus control region (LCR) or domain control region (DCR), are responsible for activation [6], and presumably decondensation, of the locus in erythroid cells. In erythroid cells, over 120 kb of DNA including and surrounding the β-globin cluster is decondensed (evidence for this is shown by the fact that it is hypersensitive to DNAase I and replicates early in the S-phase of the cell cycle), whereas in a non-expressing tissue, such as sperm, it is highly condensed [7]. However, not all transcriptionally active chromatin segments adopt an open or closed configuration as a whole: the α-globin cluster includes segments which are open even in non-expressing cells [8].

References

1. Kozak, M. (1986) *Cell*, **44**, 283.
2. Klausner, R.D. and Harford, J.B. (1989) *Science*, **246**, 870.
3. Casey, J.L., Hentze, M.W., Koeller, D.M., Wright Caughman, S.W., Roault, T.A., Klausner, R.D and Harford, J.B. (1988) *Science*, **240**, 924.
4. Hamer, D.H. (1986). *Ann. Rev. Biochem.,* **55**, 913.
5. Ptashne, M. (1986) *Nature*, **322**, 697.
6. Grosveld, F., Blom van Assendelft, B., Greaves, D.R. and Kollias, G. (1987) *Cell*, **51**, 975.
7. Townes, T.M. and Behringer, R.R. (1990) *Trends Genetics*, **6**, 219.
8. Vyas, P., Vickers, M.A., Simmons, D.L., Ayyub, H., Craddock, C.F. and Higgs, D.R. (1992) *Cell*, **69**, 781.

Further reading

Darnell, J., Lodish, H. and Baltimore, D. (1990) *Molecular Cell Biology*, 2nd edn. Scientific American Books, New York.

Lewin, B. (1990) *Genes IV*. Oxford University Press, Oxford.

Ruoslahti, E. (1988). *Ann. Rev. Biochem.* **57**, 375.

Smith, C.W.J., James, J.G. and Nadal-Ginard, B.N. (1989) *Ann. Rev. Genet.*, **23**, 527.

Strachan, T. (1992) *The Human Genome*. BIOS Scientific Publishers, Oxford.

Watson, J.D., Hopkins, N.H., Roberts, J.W., Steitz, J.A. and Weiner, A.M. (1987) *Molecular Biology of the Gene*. Benjamin Cummings. Menlo Park, CA.

2
GENE ANALYSIS TECHNIQUES. I: LOW RESOLUTION MAPPING OF GENOMIC DNA

Underpinning all the techniques used to isolate and manipulate genes (Chapter 4) are a series of methods for analyzing and characterizing DNA and RNA sequences. These are described in this and the following chapter. In this chapter we describe the classical methods of nucleic acid hybridization, their application in Southern and Northern blot analyses, and their extension to mapping DNA sequences to particular chromosomal locations and to diagnosis of human genetic diseases.

Gene structure is determined using a combination of DNA and RNA mapping techniques. Most of these methods rely to a greater or lesser extent upon the ability to perform nucleic acid hybridization.

2.1 Nucleic acid hybridization

If a double-stranded DNA molecule is exposed to high temperature, or to very alkaline conditions, then the two strands will come apart. The molecule is said to have become denatured. The temperature at which denaturation occurs is termed the melting temperature or T_m. If the denatured DNA is returned to a temperature below its T_m, or to neutral pH if alkali was used to denature it, each strand will, after a time, find its complementary strand. The two strands will 'zipper' back together to re-form a double-stranded DNA molecule. This ability of complementary sequences to anneal, or hybridize, to one another can be used to identify molecules which contain the same sequence of nucleotides. Hybridization works even when the complementary sequences form part of a complex mixture of nucleic acid molecules.

When employed analytically, hybridization is normally performed using one labeled sequence, termed the probe, and an unlabeled sequence called the target. The probe is labeled by incorporation either of radioactively labeled nucleotides or of nucleotides which are chemically modified so that they can be identified immunologically. The probe is the known, pure species in the hybridization and the target is the unknown species to be identified.

The target will most often form part of a mixture of unrelated nucleic acid sequences. The probe is usually added in considerable excess over the target, and its concentration determines the rate of reaction. Hybridization reactions occur with logarithmic kinetics, so that most of the annealing occurs early during the incubation. However, extended hybridization times, up to one or two days, are often used to ensure complete reaction. The target and probe can be hybridized to each other in solution, or alternatively the target can be immobilized on an inert support such as nitrocellulose (*Figure 2.1*). When the target is immobilized the rate of annealing for a given amount of target and probe is greatly reduced relative to that observed when both partners are free to diffuse in solution.

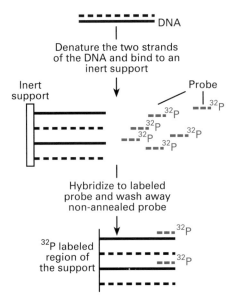

Figure 2.1: *The principle of hybridization using an inert support. The target DNA is most often localized to one part of the support and, in most cases, the aim of the hybridization experiment is to identify this region. The probe is prelabeled in some way, most often by the incorporation of radioactivity. After hybridization, the annealed probe is detected by autoradiography.*

The rate of hybridization depends upon the DNA concentration, the concentration of salt in the solution and the temperature. Hybridization reactions are normally performed at 10–20°C below the T_m of the duplex. Under any given conditions of ionic strength and temperature, the T_m of a duplex depends on the base composition of the annealed partners. AT base pairs form only two inter-strand hydrogen bonds, while GC base pairs form three hydrogen bonds. Hence, the higher the content of GC base pairs within a duplex, the higher will be its T_m. These relationships are described by the following formula for a DNA:DNA duplex:

$$T_m = 16.5 \, (\log Na^+) + 0.41 \, (\%GC) + 81.5°C,$$

where Na^+ is the sodium ion concentration and (% GC) is the percentage of GC residues in the duplex.

This relationship holds only for DNA:DNA duplexes. Duplexes in which RNA is one of the annealed partners melt at a higher temperature. DNA:DNA duplexes are less stable (lower T_m) than DNA:RNA duplexes, which in turn are less stable than RNA:RNA duplexes. At high temperatures (e.g. 70°C) RNA is degraded, so hybridizations involving RNA are performed in the presence of formamide. Formamide disrupts hydrogen bonding and thus reduces the effective T_m of the duplex, allowing annealing to occur at a lower temperature (e.g. 37°C).

The target and the probe need not be identical over their whole lengths. Non-identical sequences can cross-hybridize if they are sufficiently closely related. The effect of non-complementary base pairs, or mismatches, is to reduce the T_m. So, by controlling the temperature and concentrations of salt and formamide, it is possible to favor or disfavor cross-hybridization. The combination of these three factors is used to dictate the stringency of the hybridization reaction. For sequences shorter than 100–200 nucleotides, the T_m also depends upon the length of overlap between the two sequences, and the effect of mismatches becomes greater the smaller the overlap.

Short oligonucleotides (15–50 nucleotides long) are very frequently used as probes. By controlling the stringency of the reaction, it is possible to control the hybridization of such oligonucleotides to their target with exquisite sensitivity. For short regions of complementarity, such as are obtained using oligonucleotide probes, the mathematical relationship given above breaks down. Here the '2+4' rule is used. At an ionic strength of 1.08 M (the ionic strength of 6 × SSC, a standard buffer used in nucleic acid hybridization) each AT base pair contributes 2°C of stabilization to the duplex and each GC base pair 4°C. Hence a nucleotide sequence containing ten AT and ten GC base pairs has a T_m of 60°C.

2.2 Gel electrophoresis and detection of nucleic acids by Southern and Northern blotting

2.2.1 Gel electrophoresis

Nucleic acid molecules can be distinguished by differences in their chain length, and most methods used to study gene organization and expression combine nucleic acid hybridization with a chain length measurement.

Because they have a net negative charge at neutral pH, DNA and RNA molecules migrate towards the positive terminal when placed in an electric field (*Figure 2.2*). This process of electrophoresis is performed in an agarose or polyacrylamide gel. Nucleic acids are loaded into slots in the gel and allowed to migrate towards the positive terminal. The pores in the gel act to sieve the molecules, so that the mobility of a particular nucleic acid species depends on its length. All the molecules of a particular size move at approximately the same rate through the gel, forming a band which gradually increases in width during electrophoresis because of diffusion. The size of the molecules in any band upon the gel can be deduced by running DNA molecules of known size in a parallel track on the gel, and determining their migration position relative to the species of unknown size.

Polyacrylamide has a smaller average pore size than agarose and so is effective at separating molecules of 10–1000 nucleotides in length (1000 nucleotides is termed 1 kilobase or 1 kb), with very high resolution. The larger pores in agarose allow it to resolve much bigger molecules, up to 100 kb in length. However, the separated molecules migrate as a wider band so that the resolution (i.e. the ability to separate molecules which differ only slightly in size) is lower with agarose than with polyacrylamide. In pulsed-field gel electrophoresis (PFGE) the electrical field across the gel is repeatedly switched back and

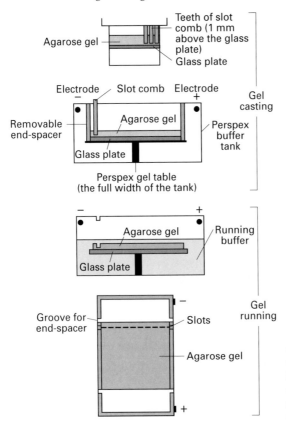

Figure 2.2: *Agarose gel electrophoresis. Redrawn from Williams and Patient (1989)* Genetic Engineering, *p. 46, with permission from Oxford University Press.*

forth between two or more directions. This allows very long DNA molecules, several million base pairs in length, to be separated, since very big molecules change direction less rapidly as they move through the agarose gel.

After gel electrophoresis, the nucleic acid molecules must be detected. They can be visualized directly by including ethidium bromide in the polyacrylamide or agarose gel. This dye binds to double-stranded nucleic acids by intercalating between adjacent base pairs, and emits a red fluorescence when exposed to UV radiation. As little as 5–10 ng of DNA can be detected (*Figure 2.3*). If the nucleic acid molecules are radioactively labeled, then individual resolved species can be detected by autoradiography. Here the sensitivity is dictated largely by one's patience in waiting for the autoradiogram to expose but an amount of ^{32}P which gives only a few disintegrations per minute (d.p.m.) within a band can readily be detected.

2.2.2 Southern and Northern blotting

Sometimes, a complex mixture of unlabeled molecules will be subjected to electrophoresis with the aim of determining the position of migration of just one or a few molecular species. This can be achieved provided that there is a hybridization probe which recognizes the specific molecule(s) of interest. In Southern blotting [1], DNA fragments are separated by agarose gel electro-

Stained Gel

1   2   3   4   5   6   7   8   9   10

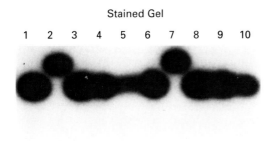

Autoradiogram

1   2   3   4   5   6   7   8   9   10

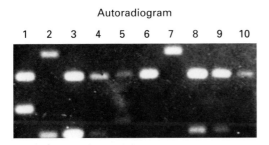

Figure 2.3: *Analysis of DNA organization using gel electrophoresis. A DNA molecule was fragmented using ten different restriction enzymes and the digestion products were subjected to electrophoresis through an agarose gel containing ethidium bromide. The gel was placed on a strong source of UV light and the red fluorescence of the ethidium bromide intercalated into the DNA was recorded photographically. The gel was then subjected to Southern blotting (see* Figure 2.4*). The DNA molecule contains an actin gene and the subset of restriction fragments which contain actin DNA sequences were detected using an actin hybridization probe. The identity of the hybridizing bands can be deduced by comparing the two panels.*

phoresis and the double-stranded DNA is denatured by incubating the gel in alkali. After neutralization of the alkali, the gel is placed in contact with a reservoir containing a buffered salt solution and overlaid with a filter which is in turn overlaid with dry paper towels (*Figure 2.4*). Buffer from the reservoir flows through the gel, carrying with it the DNA fragments. Single-stranded nucleic acids bind irreversibly to nitrocellulose filters, or can be cross-linked to other, less fragile, supports (such as nylon membranes) by exposure to UV light. During the transfer, DNA fragments retain their positions relative to one another with remarkably little loss of resolution. Northern blotting [2] is a similar process applied to RNA molecules. Electrophoresis is performed in the presence of a denaturing agent to prevent inter- and intra-molecular interactions that would prejudice separation of the different RNA species.

The membrane or filter bearing the nucleic acids is exposed to a complementary probe under appropriate hybridization conditions. Probes can be

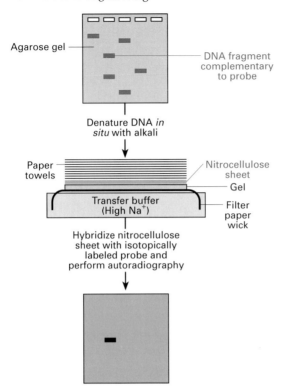

Agarose gel

DNA fragment complementary to probe

Denature DNA *in situ* with alkali

Paper towels

Nitrocellulose sheet

Gel

Transfer buffer (High Na⁺)

Filter paper wick

Hybridize nitrocellulose sheet with isotopically labeled probe and perform autoradiography

Figure 2.4: *Southern blotting.*

either cloned natural DNA or short chemically synthesized oligonucleotides. Such oligonucleotide probes, 20–50 nucleotides in length, contain the complement of the precise piece of nucleic acid which is to be detected. Chemical DNA synthesis is performed by coupling the first base, the base at the 3′ end of the desired sequence, to an inert resin and performing serial cycles of base addition. In contrast to nucleic acid polymerization reactions in nature, the chain grows in a 3′ to 5′ direction. The synthesis is usually automated using a machine which can synthesize chains up to 100 nucleotides long in a matter of hours. Whole genes 1000 or more nucleotides in length have been constructed by ligating together overlapping, chemically synthesized chains.

The probe is labeled, prior to hybridization, by incorporation of nucleotides that are either radiolabeled or are chemically modified so as to be detectable immunologically. Preparation of probes carrying highly sensitive and specific labels is crucial to the success of Southern or Northern analysis.

2.2.3 Preparation of labeled probes

For synthetic oligonucleotides, a ^{32}P isotopic label can be added to the 5′ hydroxyl group using polynucleotide kinase (*Table 2.1*). Alternatively, labeled nucleotides can be incorporated at the 3′ end using terminal transferase. Because label is added at only one place within the oligonucleotide, such end-labeled probes contain relatively little radioactivity per unit weight of DNA. This low specific activity limits their sensitivity as probes. If a high

Table 2.1: *Methods of* in vitro *labeling of nucleic acids for Southern and Northern transfer*

Labeling method	Enzyme used	Template	Product	Structure at end of reaction*	Specific activity
5' end labeling	Polynucleotide kinase	DNA or RNA	DNA or RNA	d.s. or s.s.	Low
3' end labeling	Large (Klenow) subfragment of E. coli DNA polymerase-1	DNA	DNA	d.s.	Low
3' end labeling	Terminal transferase	DNA or RNA	DNA or RNA	s.s.	Low
Nick translation	E. coli DNA polymerase-1	DNA	DNA	d.s.	High
Random priming	Klenow subfragment of DNA polymerase-1	DNA or RNA	DNA	d.s.	High
In-vitro transcription	Bacteriophage (e.g. T7) RNA polymerase	DNA	RNA	s.s.	High

* d.s. – double stranded; s.s. – single stranded.

specific activity probe is required, then some method of uniform labeling must be used. Here the probe sequence is copied along part or all of its length using DNA polymerase and labeled nucleoside triphosphates.

DNA polymerase works by extending a short double-stranded region made by annealing an oligonucleotide primer to the single-stranded template. Thus this method of uniform labeling requires a primer which matches the probe sequence. If the probe sequence is not known (as is often the case when natural cloned DNA is used), random oligonucleotide labeling [3] can be used. A large excess of short oligonucleotides, hexamers for example, of random DNA sequence is used as primer (*Table 2.1, Figure 2.5*). These are made by adding a mixture of all four bases at each step in the chemical synthesis reaction. The DNA is denatured and the two complementary strands are copied in the presence of labeled, or chemically modified, nucleoside triphosphates. The polymerase used is a proteolytic fragment (Klenow fragment) derived from DNA polymerase-1 of *E.coli*. Chance homology ensures that the random oligonucleotides anneal to the separated DNA strands at many points along their length, so providing the primers that the polymerase needs for the initiation of DNA synthesis. This is only one of several uniform labeling methods (*Table 2.1*), one of which involves an initial genetic engineering step (see Section 4.2.4).

If the probe is double-stranded DNA, a fraction of the probe will become unavailable for hybridization to the target because the two strands of the probe will anneal together. However, there will normally be a sufficient excess of

DNA to be used as a probe

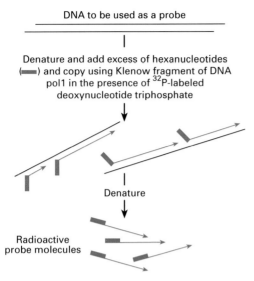

Denature and add excess of hexanucleotides
(━) and copy using Klenow fragment of DNA
pol1 in the presence of ^{32}P-labeled
deoxynucleotide triphosphate

Denature

Radioactive
probe molecules

Figure 2.5: *The principle of random primed (oligo-) labeling. The DNA to be used as a probe is denatured by heating and mixed with hexanucleotides of random sequence which then act as primers for DNA synthesis.*

probe to allow some of the probe to hybridize to the target. At the end of the hybridization, unbound probe is removed by washing. Isotopically labeled probe, bound to its target, is then detected by autoradiography (*Figure 2.6* and *Table 2.2*). Using such methods, tiny amounts of DNA, in the order of 1 pg per band, can be detected. Until recently radioactive labeling of the probe was the most sensitive detection technique available; now, however, immunodetection procedures, based upon amplification of the signal with enzymes bound to antibodies, are claimed to be equally sensitive (*Figure 2.7*). The great advantage of non-radioactive probes is that they can be stored for long periods of time because they are not subject to radioactive decay and they require no special safety precautions.

The end product of the Southern and Northern transfer is an image of the original gel in which only a subset of the products is visualized. The position of migration can be used to deduce the size of the RNA or DNA molecules hybridizing to the probe, and the strength of the signal is proportional to the amount of nucleic acid within a particular band. Therefore, by analyzing RNA

Table 2.2: *Hybridization of DNA from human–mouse hybrids with ERBA1-specific probes*

Hybrids	Result	1	2	3	4	5	6	7	8	9	10	11	12	13	14	15	16	17	18	19	20	21	22	X
DUR4.3	+	−	−	+	−	+	−	−	−	−	+	+	+	+	+	+	−	+	+	−	+	+	+	+
CTP34B4	+	+	+	+	−	+	+	+	+	+	−	−	+	−	+	−	+	+	+	−	−	−	−	+
3W4C15	+	−	−	−	−	−	+	−	−	+	+	+	−	+	+	−	+	−	−	−	+	−	+	+
CTP41P1	+	−	−	+	−	−	+	+	−	−	−	−	−	+	−	+	+	−	+	−	−	+	−	−
PotB2/B2	+	−	−	−	−	−	−	−	−	−	−	+	−	−	+	−	−	+	−	−	+	+	−	−
PCTB/A1.8	+	−	−	−	−	−	−	−	−	−	−	−	−	−	+	−	−	−	−	−	−	−	−	−
Horp27RC14	−	−	−	−	+	−	−	+	−	−	+	+	+	−	+	+	−	−	−	−	−	+	−	+

or DNA samples extracted from different sources in adjacent slots, a quantitative comparison can be made of the amount of a specific target molecule in each sample.

2.3 Mapping genes to chromosomes

One of the most important applications of nucleic acid hybridization is in the construction of genetic maps, that is, in determining the chromosomal location of individual genes. There are three commonly used methods that differ in the degree of resolution they afford.

2.3.1 Somatic cell hybrids

Somatic cell hybrids are prepared by fusing together human and rodent cells using reagents, such as polyethylene glycol or Sendai virus, which cause membrane fusion. The resulting hybrid cells are then grown in culture medium

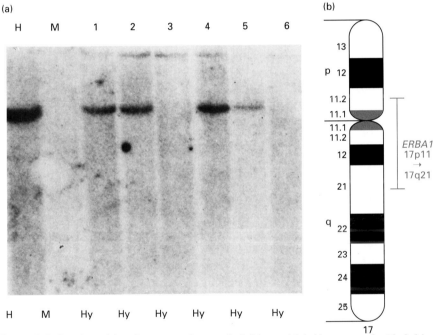

Figure 2.6: *Southern blot of a mouse–human hybrid panel (**a**). Human–rodent hybrids containing different combinations of human chromosomes can be used to map genes and proteins to specific chromosomes. Because human chromosomes can be lost during continuing culture, the hybrids must be karyotypically characterized every time a batch of DNA is prepared. In this experiment DNA from such a panel was subjected to restriction enzyme digestion and Southern blotting using a probe containing the human ERBA1 gene. Lanes H and M contain control human and mouse DNA, respectively. Lanes 1 to 6 contain DNA derived from six different mouse–human hybrids. These results show that four of the hybrids (1, 2, 4 and 5) contain the human ERBA1 gene and two (3 and 6) do not. Further analysis (Table 2.2) of a larger panel of hybrids, one of which contained a translocation of a part of human chromosome 17 allowed mapping of the ERBA1 gene to the centromere proximal region of chromosome 17. (**b**) Representation of human chromosome17 showing location of the ERBA1 gene as deduced from the mapping information in (**a**).*

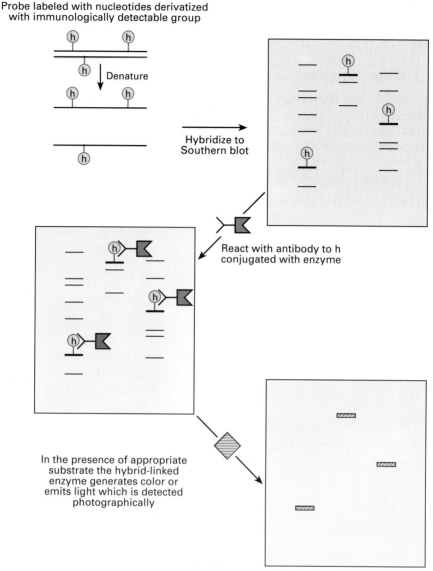

Probe labeled with nucleotides derivatized with immunologically detectable group

Denature

Hybridize to Southern blot

React with antibody to h conjugated with enzyme

In the presence of appropriate substrate the hybrid-linked enzyme generates color or emits light which is detected photographically

Figure 2.7: The principle of non-radioactive detection methods for in-situ hybridization.

containing selective agents which allow only the hybrid cells to grow. In hybrids produced using human and rodent cells, the rodent chromosomes are retained, while the human chromosomes are lost. This random chromosome loss occurs during continued culture and shows no apparent selectivity for individual human chromosomes. Thus it is possible to produce a panel of hybrid cells containing different combinations of human chromosomes. Genomic DNA is prepared from panels of such hybrids and used in hybridization analysis (normally a Southern transfer) with the human probe which is to be mapped. By determining which of the panel of hybrids shows a hybridization signal, and correlating this information with the karyotype of the cells, it is possible to deduce which chromosome carries the test gene (*Figure 2.6, Table 2.2*).

2.3.2 In-situ *hybridization*

In this technique a labeled DNA probe is hybridized to metaphase chromosomes and the position of hybridization is visualized at the light microscope level. Metaphase chromosomes are liberated from cells and bound to the surface of a cover slip. The DNA within the chromosomes is denatured in such a way that the individual chromosomes remain visually identifiable. The probe is labeled, either by incorporation of a ^{3}H-nucleotide or non-isotopically. The chromosomcs arc incubated with the labeled probe for many hours and, after hybridization, the excess unbound probe is washed away and the annealed probe detected.

If the probe was radiolabeled, the slides are dipped into a photographic emulsion. The radioactive emission from the bound radiolabel interacts with the emulsion, depositing a silver grain. The grains will lie in the approximate area at which the probe bound to the chromosome. The accuracy of mapping by this technique is dependent on the size of the grains which, in the case of small chromosomes, can cover a significant portion of the total chromosome length.

If the probe is labeled non-isotopically then it is possible to determine the precise position of hybridization with much greater accuracy [4]. The signal is detected using a fluorescent molecule so that there is no spread of the signal such as occurs with autoradiography. This is called fluorescent *in-situ* hybridization (FISH) (*Figure 2.8*). It is more sensitive than radiolabeling and can be used to give a fairly precise position of hybridization. Thus FISH is particularly useful for determining the respective locations of two markers on the same chromosome. The two probes are labeled with different immunologically labeled adducts and a single hybridization reaction is performed. The two probes are then detected using fluorochromes which emit at different wavelengths when excited by UV light, such as rhodamine (red fluorescence) and fluorescein (green fluorescence). This technique has been used to order probes along chromosomes when they are separated by no more than a few hundred kilobases of DNA. The FISH technique can also be applied to interphase nuclei, which allows accurate ordering of probes along a chromosome when they are separated by no more than 20–30 kilobases.

2.3.3 Genetic linkage mapping

This form of mapping is used to construct genetic linkage maps of individual chromosomes and also to localize genes involved in genetic diseases. Genetic linkage analysis relies on an ability to estimate the frequency of crossing over (recombination) that occurs between homologous chromosomes during meiosis. The closer two genetic markers are to each other then the lower is the chance that they will be separated during meiosis. The statistical probability that two markers are linked is measured as the logarithm of the odds (Lod). A

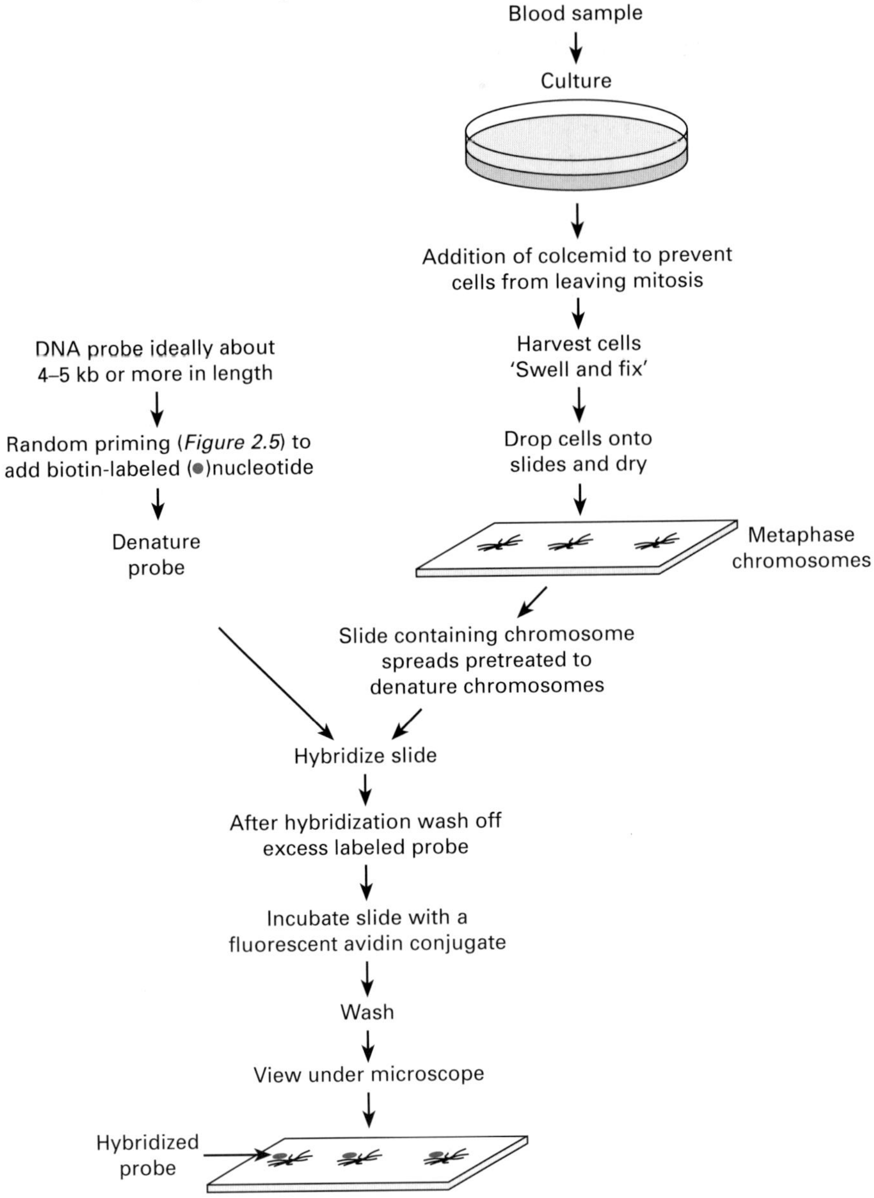

Blood sample

↓

Culture

↓

Addition of colcemid to prevent
cells from leaving mitosis

↓

DNA probe ideally about
4–5 kb or more in length

↓

Random priming (*Figure 2.5*) to
add biotin-labeled (●)nucleotide

↓

Denature
probe

Harvest cells
'Swell and fix'

↓

Drop cells onto
slides and dry

↓

Metaphase
chromosomes

Slide containing chromosome
spreads pretreated to
denature chromosomes

↓

Hybridize slide

↓

After hybridization wash off
excess labeled probe

↓

Incubate slide with a
fluorescent avidin conjugate

↓

Wash

↓

View under microscope

↓

Hybridized
probe

Figure 2.8: *FISH is becoming the technique of choice in gene mapping experiments using* in-situ *hybridization. It uses a non-radioactive labeling system which involves the incorporation of nucleotides with molecules of biotin attached. These have a strong affinity for avidin which in this example has been conjugated with a flourescent molecule. When this is viewed under special filters the specific sites to which the DNA probe has bound can be seen on a chromosome.*

Lod score of 3 is usually taken as the threshold of significance, while a Lod score below −2 suggests that the two loci are not linked. Lods between −2 and +3 are not significant.

2.4 Restriction enzyme mapping and its applications

The above three techniques allow a gene to be mapped to a specific chromosome, and linkage analysis and *in-situ* hybridization allow its approximate position to be deduced relative to a known marker. However, more complex forms of chromosome analysis, which allow for the precise ordering of individual genes, are dependent on the ability to fragment the chromosomal DNA.

2.4.1 Restriction enzymes

Many of the great advances in molecular biology have come about because of the discovery of enzymes which cleave double-stranded DNA into discrete pieces, resolvable by gel electrophoresis. Restriction enzymes are prokaryotic proteins which probably exist as a defense against infection by bacterial viruses. They cleave at, or very near to, specific recognition sequences within DNA. They are named after the species from which they derive (e.g. the enzyme *Hin*dIII derives from *Hemophilus influenzae*). Each species of bacterium contains a methylase which adds methyl groups to its DNA, protecting it from cleavage by its own restriction enzyme. Incoming foreign DNA lacks this modification and is destroyed by the restriction enzyme.

The restriction enzymes in common use recognize palindromic sequences within their DNA target. When read in a 5′ to 3′ direction, the sequence in each of the two complementary strands is identical (*Figure 2.9*). The recognition

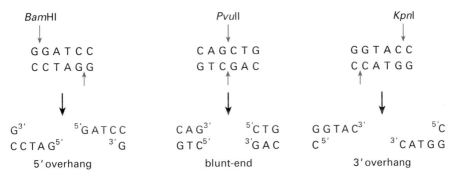

Figure 2.9: *The three classes of cleavage sites in commonly used restriction enzymes.*

sequence is normally four, five or six nucleotides in length. The length and base composition of the recognition sequence determine its frequency of occurrence within any particular piece of DNA. Thus the four nucleotide recognition sequence for the enzyme *Sau*3A, GATC, will occur once every 256 (4^4) nucleotides in DNA which contains equimolar amounts of the four bases. The six nucleotide recognition sequence for *Bam*HI, GGATCC, will occur once every 4096 nucleotides. Since they have the same four central residues, GATC, *Bam*HI will cut at a subset of *Sau*3A sites.

The frequency with which most restriction enzymes will cut a target DNA is predictable from the base composition of the target and the length and nucleotide sequence of the recognition site. However, certain cleavage sites are greatly under-represented in genomic DNA from mammalian cells. These are sites which include the doublet CpG, for example, the recognition site for the enzyme *Sma*I which cuts at the sequence CCCGGG. The CpG dinucleotide is under-represented because it is a substrate for DNA methylation. Not all CpG dinucleotides within a particular tissue are modified and there is great variation between tissues in the extent of methylation at a particular CpG doublet. In general, a gene is under-methylated in those tissues in which it is expressed. Thus in sperm cells the region of genomic DNA containing the globin genes is heavily methylated while in erythroid cells it is much less highly modified [5].

When cytosine in a CpG dinucleotide is methylated to form 5-methyl cytosine (*Figure 2.10*) this has two consequences. First, many enzymes will not cut a recognition site which is methylated in this way. Secondly, over evolution-

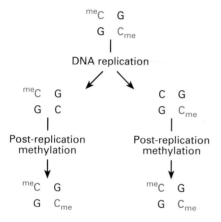

Figure 2.10: *Inheritance of methylation at CpGs. This represents part of a double-stranded DNA molecule containing a CpG dinucleotide in which the C residue is methylated. When replicated, the two daughter molecules are initially methylated only in the parental strand (they are said to be hemi-methylated). A maintenance methylase then acts to modify the C residue in the newly synthesized strand.*

ary time 5-methyl cytosine tends to be lost by spontaneous deamination. Unmethylated cytosine also deaminates, but the product is uracil, an unnatural base in DNA, which is recognized and repaired by the cell. 5-methyl cytosine however, deaminates to the natural base thymine, which is therefore not repaired. Thus mammalian genomic DNA contains far fewer CpG sequences than its overall base composition would predict, and many of those that are present are methylated. Because of this, enzymes whose recognition site includes one or more CpG sequences are known as rare-cutters. The enzyme *Not*I has an eight nucleotide recognition sequence GCGGCCGC with two CpG dinucleotides, and so it cleaves very infrequently in mammalian DNA. Such rare-cutter enzymes are particularly useful for creating large fragments containing long segments of genomic DNA for analysis by PFGE.

Some enzymes cut DNA at the same position in each strand, to generate flush or blunt termini (*Figure 2.9*). Others cut in an offset manner, to yield staggered termini. Since these are complementary, and hence mutually cohesive, they are often termed sticky ends.

Any piece of DNA of sufficient length will have recognition sites for one or more restriction enzymes and it is possible to determine the position of these sites. This is done by digesting the DNA with single enzymes, and with combinations of two enzymes (this is termed a double digest), and determining the sizes of the resultant fragments by gel electrophoresis (*Figure 2.11*). With a sufficient number of digests, it is possible to deduce the origin of each band on the gel, that is, which enzyme or combination of enzymes generated it. This

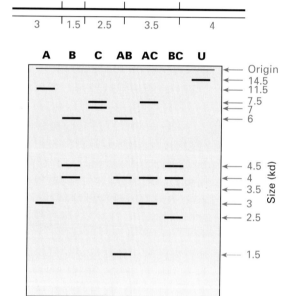

Figure 2.11: *Restriction mapping of a discrete piece of DNA. The DNA is shown in the upper part of the figure as a linear fragment containing restriction sites A, B and C, with two sites for enzyme B. By performing double digests with all combinations of the three enzymes, it is possible to deduce the relative positions of the cleavage sites and hence deduce the map. U is undigested DNA.*

yields a restriction enzyme cleavage map, more commonly called a restriction map. Obtaining such a map is very often an essential first step in analyzing a piece of DNA of unknown structure.

2.4.2 Restriction enzyme mapping of genomic DNA

If total human genomic DNA is cut with a restriction enzyme such as *Eco*RI, which cuts on average about once every 4000 nucleotides, then approximately one million fragments are generated. When analyzed by agarose gel electrophoresis and stained with ethidium bromide, such a digest yields a smear extending throughout the gel. However, any individual *Eco*RI fragment will be of a specific size and it will therefore migrate as a single band. It is possible to detect such a band, by performing a Southern transfer and hybridizing the filter with a probe complementary to some part of the fragment. Again, a restriction map of sites around the fragment can be built up by performing multiple digests of the total DNA. Here, however, in contrast to the situation described above for a single piece of DNA, it is only possible to identify the sites immediately flanking the probe (*Figure 2.12*).

2.4.3 Diagnosis of human genetic diseases by restriction fragment length polymorphisms

Another major use of Southern transfer analysis is in diagnosing human genetic diseases. Major alterations in chromosome structure such as deletions of 10 Mb or more of DNA or translocation between chromosomes can be detected by karyotyping. However, most genetic diseases are caused by less dramatic alterations which must be localized by gene mapping before diagnosis can be performed.

There are four copies of the α-globin gene per diploid human genome, with two highly homologous genes present in a tandem orientation on chromosome 16. In the homozygous lethal form of α-thalassemia all four genes are deleted

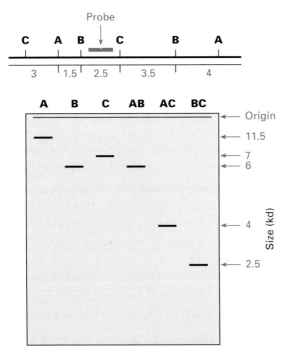

Figure 2.12: *Restriction mapping by Southern transfer analysis of a piece of DNA forming part of a complex mixture of restriction fragments. In contrast to the situation described in Figure 2.11, where a discrete piece of DNA is analyzed on an ethidium bromide-stained gel, the DNA here is analyzed by Southern transfer using fragment B–C as a probe. This is the situation that arises when an entire mammalian genome is being analyzed to map a single gene from within it. It is only possible to detect the subset of fragments generated in the double digestion reactions which lie adjacent to the probe. For example, fragment C–A (at the extreme left-hand side of the map) will not be detected while the adjacent fragment A–C will be detected. Thus, for any one enzyme, the deduced map only gives information as to the nearest cleavage site upstream or downstream of a particular probe (i.e. an A site within the upstream C–A fragment would not be detected).*

and such fetuses normally die during pregnancy, or a few days after birth. Homozygous deletion of the four α-globin genes can easily be diagnosed by probing a Southern blot with a probe that recognizes the α-globin gene. The copy number of the gene can also be determined, so that heterozygous individuals can be identified.

The most favorable situation for diagnosis of point mutations, or deletions, or insertions that are too small to be detected as a size shift on a gel, is a sequence change that creates a new restriction site or destroys an existing one. In sickle cell anemia a point mutation in the β-globin gene results in the glutamic acid at position 6 being changed to a valine residue [6]. The normal sequence at this point is CCTGAGGAG, while the mutated sequence is CCTGTGGAG. The restriction enzyme *Mst*II recognizes the sequence CCTNAGG, where N is any one of the four nucleotides. *Mst*II will therefore digest the β-globin gene at this position, while there will be no equivalent cut in the mutant gene. Such a change can readily be detected (*Figure 2.13*) and

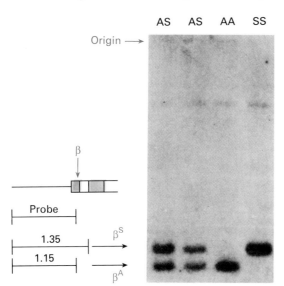

Figure 2.13: *The diagnosis of sickle cell anemia by Southern blot analysis. The sickle cell mutation disrupts the* MstII *site encompassing codons 5 – 7 of the β-globin gene. The 1.35 kb band resulting from the loss of the* MstII *site is detected on a Southern transfer. AS: heterozygous individuals (carriers); AA: normal individual (homozygous); SS: affected individual (homozygous for the sickle cell gene). Reproduced from Marx (1989) with permission from Cambridge University Press.*

sufficient DNA for Southern transfer analysis can be isolated by chorion villus biopsy or amniocentesis, so that this test can be be used for prenatal diagnosis. A similar form of diagnosis can be performed for deficiency of the clotting factor VIII [7], where several of the mutations that affect the gene also destroy restriction sites.

The point mutation that causes sickle cell anemia was discovered because it has such a disastrous effect upon the afflicted individual. The fact that the point mutation causing sickle cell anemia changes the restriction map of the β-globin gene is fortuitous, but a detectable change of this kind is the exception rather than the rule (because most point mutations will be within regions of DNA sequence that do not contain the cleavage site for any restriction enzyme). Fortunately, there is an alternative method of diagnosing such changes.

Analysis of human DNA from different individuals has revealed large numbers of sequence differences that produce no obvious phenotypic effects, often because they lie between genes or within introns. When one of these cryptic or silent changes is frequent in a population it is called a polymorphism. A change that produces a difference in the restriction map of a region is called a restriction fragment length polymorphism (RFLP). The degree of polymorphism between two individuals depends upon their ethnic relationship. Even within a single population, however, there is likely to be variation within the restriction map encompassing a particular gene. A good example of the extent of natural variation which exists is provided by the map of the β-globin gene cluster in the Italian population (*Figure 2.14*). If a polymorphism can be found that is closely linked to a gene responsible for a genetic disease, then it can be used for prenatal diagnosis.

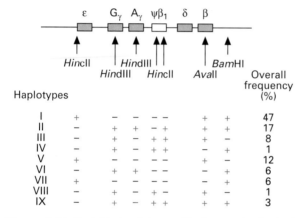

Haplotypes	HincII	HindIII	HindIII	HincII	AvaII	BamHI		Overall frequency (%)
I	+	−	−	− −	+		+	47
II	−	+	+	− +	+		+	17
III	−	+	−	+ +	+		−	8
IV	−	+	−	+ +	−		+	1
V	+	−	−	− −	+		−	12
VI	−	+	+	− −	−		+	6
VII	+	−	−	− −	−		+	6
VIII	−	+	−	+ −	+		−	1
IX	−	+	−	+ +	+		+	3

Figure 2.14: *β-globin haplotypes. Haplotypes in the Italian population deduced from seven restriction sites in the β-globin gene cluster. The first line shows the gene arrangement in the β-globin gene cluster, with the position of sites for seven restriction enzymes. The table below illustrates the patterns and the frequencies of the nine haplotypes found in the Italian population. The symbols + and − indicate the presence or absence of a particular restriction site.*

An instructive example of the use of this approach in diagnosing disease is again provided by the β-globin gene. In most genomes carrying the sickle cell mutation there is a second change, additional to the missing *Mst*II site [8]. An *Hpa*I site, located outside the coding region of the gene, is absent so that *Hpa*I digestion yields a 13 kb fragment rather than the normal 7.6 kb fragment. These two differences from the normal human genome, the missing *Hpa*II and *Mst*II sites, are often found to reside together on the same chromosome. Because the sites are tightly linked, so that recombination rarely separates them at meiosis, the two variants tend to remain associated from generation to generation.

By performing a pedigree analysis of an individual suspected of carrying the sickle cell anemia gene, it is possible to correlate the presence of the 13 kb *Hpa*I fragment with the sickle cell trait in his or her antecedents (*Figure 2.15*). If such a correlation exists, and if the individual is homozygous for the 13 kb fragment,

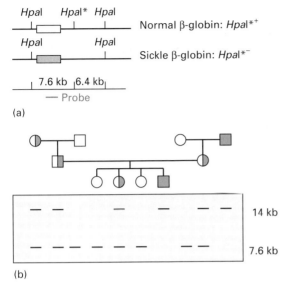

Figure 2.15: *Pedigree analysis of a family with sickle cell anemia. (**a**) The positions of Hpal sites around the β-globin gene are shown, with the sizes of the fragments indicated in kb, and the region used as a probe. Hpal* is the diagnostic site. (**b**) Scheme of a Southern blot of an Hpal genomic digest. Each lane corresponds to the member shown above it in the pedigree. The sizes of the fragments are indicated on the right. ○ – female; □ – male. Filled symbol – Hpal*⁻/ Hpal*⁻; half-filled symbol – Hpal*⁺/Hpal*⁻; open symbol – Hpal*⁺/Hpal*⁺.*

he or she is highly likely to be affected by the disease. The only circumstance under which this will not be true is if there has been a meiotic recombination between the *Hpa*I site and the mutant β-globin gene in one of the germ cells that gave rise to the individual. Depending upon the nature of the recombination event, the individual could possess a normal gene but be incorrectly diagnosed as being homozygous for the sickle cell gene or vice versa. If, however, the linkage of the polymorphic site to the genetic lesion is sufficiently close, such recombinational events will be very rare. Thus the *Hpa*I site is a very powerful indicator of the disease but is not as totally predictive as the *Mst*II site, and it requires that pedigree analysis be performed.

The great advantage of RFLP mapping is that it is an essentially universal method. This is well illustrated by the β-thalassemias. There are more than 50 different mutations that result in loss of expression of the β-globin gene. Some rare forms result from large deletions of the gene, but the most frequent lesions are point mutations. More than 30 different alleles have been described. Here the only realistic possibility is to diagnose the presence of the β-thalassemic trait by RFLP analysis. At the time of writing the RFLP-based diagnostic tests for phenylketonuria, α1-antitrypsin deficiency and clotting factor VIII and IX deficiencies have been superseded by direct identification of the most frequent mutations using a synthetic oligonucleotide probe. However, RFLP analysis is the only available method for prenatal diagnosis of those diseases in which the chromosomal location is known but the gene has not yet been identified. At the time of writing this is true for Huntington disease and adult polycystic kidney disease.

2.4.4 *Mapping variable number tandem repeats*

In addition to the simple polymorphisms detected by the presence or absence of a restriction enzyme cleavage site, variable number tandem repeats (VNTRs) provide another major source of polymorphic variation in humans

[9]. VNTRs are short tandemly repeated DNA sequences, where the number of repeat units at a particular chromosomal location varies from person to person. The size of different repeated units varies from a single nucleotide up to tens of nucleotides. Repeats of 1–4 nucleotides are called microsatellites; their special features are discussed in Chapter 3.

The major advantage of VNTRs over RFLPs is that VNTR markers show multi-allelic variation. Rather than just two alleles, giving rise to only three possible combinations of alleles in a lineage (*Figure 2.15*), there is great variation in the size of each VNTR within the genome of different individuals (*Figure 2.16*). This high level of individual variation is the basis for their use in

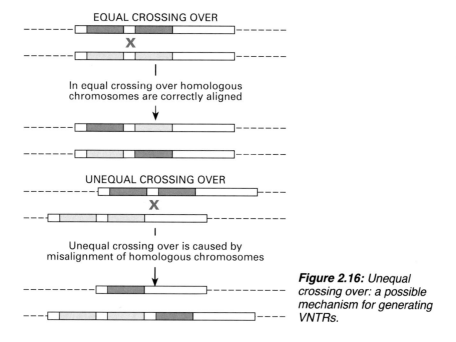

Figure 2.16: Unequal crossing over: a possible mechanism for generating VNTRs.

paternity and identity testing in forensic medicine. When a combination of four or five VNTR markers is used they can confirm identity with a high level of certainty.

The cause of the high level of allelic variation between VNTRs has been controversial. One theory ascribes it to unequal crossing over. At meiosis the pairs of chromosomes come together and recombination leads to an exchange of genetic material between homologs. If, during this crossing over, the chromosomes are not aligned perfectly, there will be a non-reciprocal exchange of material. This is termed unequal crossing over (*Figure 2.16*). The presence of tandem repeats in the genome could lead to misalignment and unequal crossing over, which would perpetuate and extend the degree of allelic variation between VNTRs. Although this explanation sounds plausible, evidence for recombination has been hard to find. It now seems more likely that variants arise during DNA replication, by slippage or 'stuttering' of the polymerase.

If a Southern blot analysis is undertaken using a probe which detects a restriction fragment containing a VNTR, then different individuals in the population will display two different sized restriction fragments, one from the maternally inherited chromosome and the other from the paternally derived chromosome. Changes in the repeat number of some VNTRs occur much more frequently than other DNA sequence changes, such as point mutations or deletions of non-repetitive sequences. However, in general, highly variable VNTRs are sufficiently stable to be useful genetic markers, stably inherited through enough generations to allow a pedigree analysis (similar to that presented in *Figure 2.15*). Thus, provided that a VNTR can be found within a reasonably close distance of a gene of interest, VNTRs can be used in just the same way as conventional RFLPs to diagnose genetic diseases. The first VNTR to be described [10] was assigned to the telomeric region of the long arm of chromosome 14. Initially, VNTRs were thought to be distributed randomly throughout the genome but, as more have been accurately mapped, it has become evident that they tend to be clustered towards chromosomal telomeres.

For genetically unrelated individuals, the precise size of the two bands observed will almost always be different. Even when individuals are closely related genetically, they can be distinguished by the use of several different probes. This is not true of conventional RFLPs, because there are only a restricted number of variants within any population. Thus VNTRs provide the first true 'molecular fingerprint' which, with the exception of monozygotic twins, is unique for every member of the human race. There are two very important applications of this fact, paternity testing and forensic medicine.

If pedigree analysis is performed on an individual using many different VNTR polymorphisms, then every band (allele) seen in that person's DNA should be present in one or other of the parents. Each parent will of course also display some length variants that are not present in their child's genome, because the child inherits only half of each parent's genes. But if the child has several bands not seen in the DNA of the supposed parents, then it is highly unlikely that the supposed parentage is correct. Usually the explanation is non-paternity, but alternatives are unacknowledged adoption or babies mistakenly switched at birth. Exclusion of alleged paternity is thus straightforward: if the child shares few or no bands with the alleged father, and has bands not present in him or the mother, then the alleged father is not the true father. Thus VNTRs provide a perfect method of paternity testing and there are now commercial companies that provide such services.

With suitable VNTRs the spectrum of bands is so variable that the chance of any other individual in the population coincidentally sharing a high fraction of variants can be extremely low. This allows positive assignment of paternity with very high probability (but never 100%), something which is not possible with any other class of genetic markers.

Because the spectrum of VNTRs is unique to an individual, they also provide enormously useful tools for forensic identification. Provided only that sufficient tissue is left at the scene of a crime, then VNTR analysis can be used to distinguish between suspects. For Southern analysis sufficient tissue must be available to extract microgram quantities of DNA. This would be a serious limitation if blood or semen were to be used in the analysis. Fortunately, however, there exists a very much more sensitive analysis method called polymerase chain reaction (PCR), which can be applied to vanishingly small tissue samples. Description of this method requires a knowledge of fine structure mapping techniques, and this is the subject of the next chapter.

References

1. Southern, E.M. (1975) *J. Mol. Biol.*, **98**, 503.
2. Alwine, J.C., Kemp, D.J. and Stark, G.R. (1977) *Proc. Natl. Acad. Sci. USA*, **74**, 5350.
3. Feinberg, A.P. and Vogelstein, B. (1984) *Anal. Biochem.*, **137**, 266.
4. Trask, B.J., Massa, H., Kenwrick, S. and Gitschier, J. (1991) *Am. J. Hum. Genet.*, **48**, 1.
5. Barker, D., Schafer, M. and White, R. (1984) *Cell*, **36**, 131.
6. Kan, Y.W. and Dozy, A.M. (1978) *Lancet*, **ii**, 910.
7. Gitschier, J., Wood, W.I., Goralka, T.M., *et al.* (1985) *Nature*, **312**, 326.
8. Kan, Y.W. and Dozy, A.M. (1980) *Science*, **209**, 388.
9. Jeffreys, A.J., Wilson, V. and Thein, S.L. (1985) *Nature*, **317**, 818.
10. Wyman, A.R. and White, R. (1980). *Proc. Natl. Acad. Sci USA*, **77**, 6754.

Further reading

Anand, R.E. (1992) *Techniques for the Analysis of Complex Genomes*. Academic Press, San Diego.

Antonarakis, S.E. (1989) Diagnosis of genetic disorders at the DNA level. *New Engl. J. Med.*, **320**, 153.

Botstein, D., White R.L., Scolnick, M. and Davis, R.W. (1980). Construction of genetic linkage map in man using restriction fragment length polymorphisms. *Am. J. Hum. Genet.*, **32**, 314.

Davies, K.E. and Tilghman S.M. (1991) *Genome Analysis*, Vol. 1, *Genetic and Physical Mapping*, Cold Spring Harbor Press, New York.

Glover D.M. (1990) *DNA Cloning: A Practical Approach*. IRL Press, Oxford.

Marx, J. (1989) *A Revolution in Biotechnology*. Cambridge University Press, Cambridge.

Wetherall, D.J. (1991) *The New Genetics and Clinical Practice*, 3rd Edn. Oxford University Press, Oxford.

Winnacker, E.-L. (1989) *From Genes to Clones, Introduction to Gene Technology*. VCH Verlag, Weinheim.

3
GENE ANALYSIS TECHNIQUES II: DETERMINING THE STRUCTURE OF GENES

3.1 DNA sequence analysis

Restriction enzyme mapping and Southern blotting help establish the gross organization of a piece of DNA, and allow the approximate position of sequences of interest to be determined. The next step in the analysis is normally to determine the nucleotide sequence.

3.1.1 Sequencing by chemical degradation

Techniques for determining DNA sequence rely upon the remarkable resolving power of polyacrylamide gels run in the presence of urea as a denaturant. If two single-stranded DNA molecules, identical in sequence over most of their length but differing by the presence of just one additional base on one of the two molecules, are subjected to electrophoresis in a denaturing polyacrylamide gel, they will be separated. In the chemical degradation procedure [1], a DNA restriction fragment is isotopically labeled with ^{32}P at either its 5' or its 3' termini and each of the two strands is separated and isolated. The DNA is then partially modified with chemical reagents specific for the different bases, and cleaved at the modified nucleotides. This generates a set of molecules differing in length but with the same isotopically labeled terminus. These fragments are subjected to denaturing gel electrophoresis in parallel slots of a high resolution gel, and the sequence of the DNA can be deduced from the resulting autoradiogram (*Figure 3.1*).

3.1.2 Sequencing by chain termination

In practice the chemical degradation procedure has been almost totally replaced by the much more convenient chain termination method [2]. In this procedure DNA is synthesized rather than degraded, but the principle is similar to the chemical method. DNA polymerases function by extending a pre-existent DNA strand, the primer, which is annealed to the template strand in the configuration shown in *Figure 3.2*. In the chain termination method, a short synthetic oligonucleotide is used as the primer, to provide a fixed 5' terminus for the growing chain.

The reaction is performed in the presence of chain terminators. These are nucleotide derivatives which contain a blocking group at their 3' hydroxyl

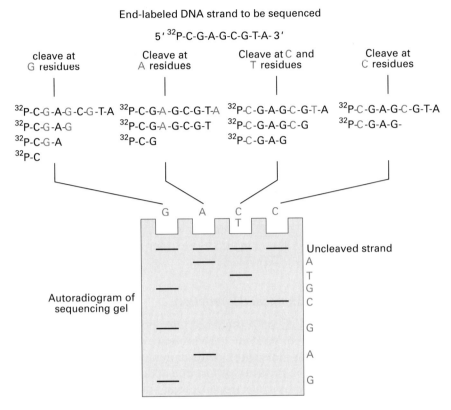

Figure 3.1: *DNA sequence analysis by the chemical degradation (Maxam and Gilbert) technique.*

group. Once incorporated into the growing chain, they prevent further extension. A small amount of chain terminator, for example, dideoxy GTP (*Figure 3.3*), is mixed with unmodified GTP and the other three nucleoside triphosphates. In this instance, a series of reaction products is produced which terminate at a subset of G residues in the newly synthesized strand (*Figure 3.2*). Four parallel reactions are run, each including a dideoxy chain terminator for one of the four nucleotides, and the products are analyzed in parallel slots of a denaturing polyacrylamide gel. The growing chain is rendered detectable by including a radiolabeled nucleoside triphosphate in the synthesis reaction. As with the chemical cleavage procedure, the nucleotide sequence of the template strand can be deduced by comparison of the pattern of autoradiographic bands in the four lanes of the gel (*Figure 3.4*).

DNA sequence analysis using chain terminators is best performed using a single-stranded template. This allows the primer free access to the template. A suitable template for single-strand DNA sequence analysis can be created by inserting the double-stranded template DNA into the genome of a bacterial virus which naturally packages only one of the two strands into viral particles (see Sections 4.2.4 and 4.3.1). Insertion of the DNA into such a vector does, however, add another step to the sequencing procedure, and this is not always

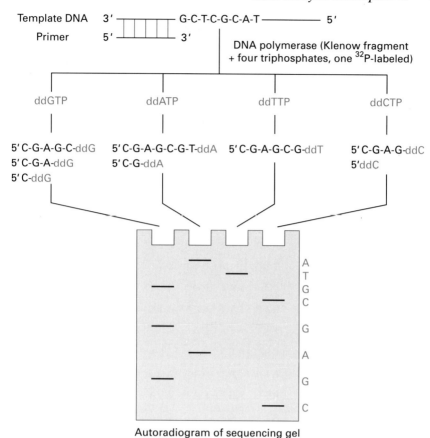

Figure 3.2: *DNA sequence analysis by the chain termination (Sanger) method.*

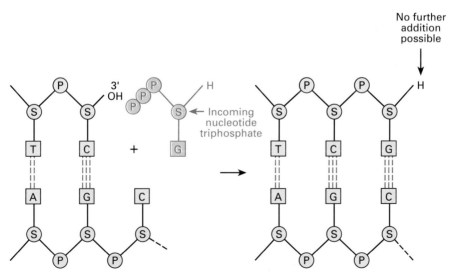

Figure 3.3: *The chemical basis of sequence analysis using chain terminators. This is a representation of a reaction in which a dideoxy G residue incorporated into the growing chain blocks further incorporation.*

convenient. If a double-stranded DNA template is denatured by treatment with alkali, and an excess of primer is added before re-annealing occurs, the primer will compete with the complementary strand for annealing to the template. This arrangement allows double-strand DNA sequence analysis to be performed, which is now very often the method of choice.

The major disadvantage of double-stranded relative to single-stranded DNA sequence analysis is that the presence of the competing complementary strand leads to a higher level of sequencing artefacts. These are positions on the gel where more than one band is present (*Figure 3.4*), so that it may be

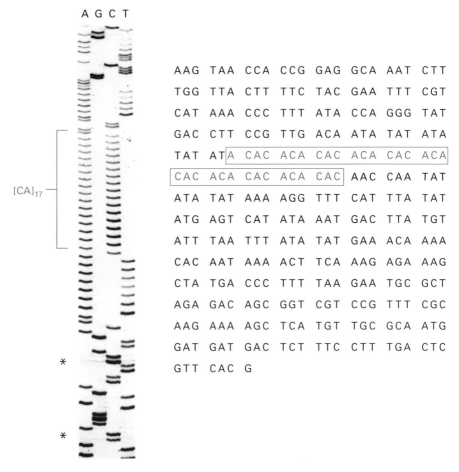

```
             A G C T

             AAG  TAA  CCA  CCG  GAG  GCA  AAT  CTT

             TGG  TTA  CTT  TTC  TAC  GAA  TTT  CGT

             CAT  AAA  CCC  TTT  ATA  CCA  GGG  TAT

             GAC  CTT  CCG  TTG  ACA  ATA  TAT  ATA

             TAT  AT A  CAC  ACA  CAC  ACA  CAC  ACA

             CAC  ACA  CAC  ACA  CAC  AAC  CAA  TAT

 [CA]₁₇      ATA  TAT  AAA  AGG  TTT  CAT  TTA  TAT

             ATG  AGT  CAT  ATA  AAT  GAC  TTA  TGT

             ATT  TAA  TTT  ATA  TAT  GAA  ACA  AAA

             CAC  AAT  AAA  ACT  TCA  AAG  AGA  AAG

             CTA  TGA  CCC  TTT  TAA  GAA  TGC  GCT

             AGA  GAC  AGC  GGT  CGT  CCG  TTT  CGC

             AAG  AAA  AGC  TCA  TGT  TGC  GCA  ATG

             GAT  GAT  GAC  TCT  TTC  CTT  TGA  CTC

    *        GTT  CAC  G
```

Figure 3.4: *Typical DNA sequence data obtained using the chain termination method. This shows the sequence of a region of a gene that contains a repeat of the sequence CA (see p. 58). Most of the sequence is completely unambiguous but the positions of some artefactual bands are indicated by *. Here there are bands in all four lanes, although in each case one band is much more intense than the other three. This band would usually be provisionally interpreted to be the 'correct' nucleotide at this point in the chain, e.g. the topmost indicated artefact would be interpreted to be a C residue. It is normally essential to determine the sequence of both strands of a DNA molecule, so that such provisional assignments would hopefully be confirmed when the other strand was sequenced.*

impossible to decide which is the correct residue at that position. Artefacts of this kind are not restricted to double-strand DNA sequence analysis; some level of artefacts is a universal problem in sequence analysis. The resultant ambiguities can often be resolved only by re-sequencing the template using another primer, or by using the other strand (*i.e.* the strand which is complementary to that used in the original analysis) as template.

Even using a single-stranded DNA as template, it is not possible on a polyacrylamide gel to resolve fragments differing by one nucleotide in length when the molecules are larger than about 500 nucleotides. This sets the upper limit to the length of sequence which can be determined using a particular primer–template complex. One way to extend from a known into an unknown region of sequence is to synthesize an oligonucleotide primer which is complementary to a tract near the end of the region whose sequence has already been established. In this way it is possible to 'walk' along a sequence, at each step synthesizing a new primer which is used to extend the region of known sequence (*Figure 3.5*). Alternatively, it is possible to delete sequences enzymatically from the template at a position downstream of the annealing site of the primer, to generate a deletion series (*Figure 3.6*). The nucleotide sequence of a set of such molecules can be used to deduce the sequence of the intact template.

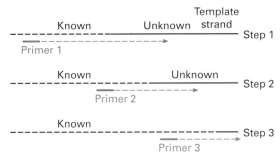

Figure 3.5: DNA sequence analysis by 'primer walking'.

By determining the sequence of both strands of a DNA molecule, and confirming that the two deduced sequences are perfectly complementary, it is possible to detect and eliminate artefacts generated in the reactions or during electrophoresis. When this check is performed, DNA sequence can be established with a very high degree of accuracy.

3.1.3 Automated DNA sequence analysis

Using the chain termination method, DNA sequence can be determined very rapidly and there are machines based upon this approach which speed the process even further (*Figure 3.7*). In a conventional sequencing reaction four separate single-strand extension reactions are performed, each containing a different chain-terminating dideoxynucleoside triphosphate. The radiolabeled reaction products are separated on four separate tracks of a polyacrylamide gel. In machine-based sequencing a single reaction is run, with each of the four dideoxynucleoside triphosphates containing a different fluorescent label.

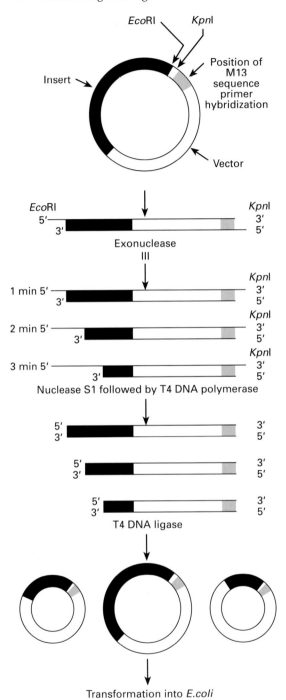

Figure 3.6: *The use of nested deletions for DNA sequence analysis.*

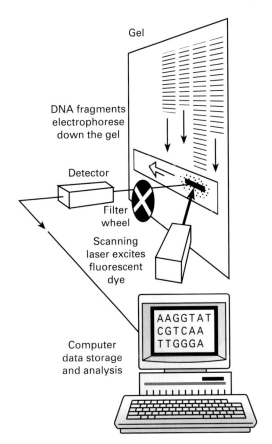

Gel

DNA fragments
electrophorese
down the gel

Detector

Filter
wheel

Scanning
laser excites
fluorescent
dye

AAGGTAT
CGTCAA
TTGGGA

Computer
data storage
and analysis

Figure 3.7: *Automated DNA sequence analysis. The chain terminator method of DNA sequencing has been modified to allow DNA sequencing to be automated. Fluorescent dyes have been developed and these are attached to each of the dideoxynucleotides which replaces the need for radioactive nucleotides. After the sequencing reactions have been completed the products of all four reactions are pooled and electrophoresed in one lane of an acrylamide sequencing gel. As the reaction products pass through the gel the specific DNA bases are detected by a scanning laser that excites each fluorescent dye at different wavelengths, this information is captured on a detector and passed to a computer for data storage and analysis.*

When these are excited by an argon laser, the fluorescent labels each emit light at a different wavelength. The same chain-termination reactions are carried out; however, it is no longer necessary to run the products in separate lanes of a gel. The reaction products are mixed and run in a single lane in a gel and as the different length fragments pass the laser each fluorescent label gives a unique signal which is stored in the computer.

Machine-based DNA sequencing is beginning to change the speed and scope of sequencing projects dramatically. The first complete genomic sequences were those of simple viruses such as the SV40 tumor virus, containing around 5000 bases of DNA. This took a number of years to complete.

Recently, the DNA sequence of over 106 000 bases of DNA from human chromosome 19 was determined in a few months using DNA sequencing machines [3]. The sequence of a significant fraction of the *E. coli* genome has been established, the entire sequence of yeast chromosome 3 is known [4], and there are plans eventually to determine the sequence of the entire human genome.

3.2 Transcript mapping techniques

Analyzing the nucleotide sequence of a piece of DNA is a necessary preliminary to understanding its precise organization and function, but is not normally sufficient for these purposes. Because eukaryotic gene transcripts are subjected to a complex series of post-transcriptional processing events, the structure of a gene can be understood only by correlating the organization of the genomic DNA with the organization of its mRNA transcript. This has required the establishment of sophisticated cloning and mapping techniques which can, when used in conjunction, give a complete description of the structure of a gene.

It is possible to isolate, by gene cloning, the double-stranded DNA copies of individual mRNA sequences (see Chapter 4). These cDNA clones will normally contain a DNA copy of the poly(A) sequence at the 3′ terminus. Hence, by determining which strand of the cDNA clone contains a poly(A) tract at one end, it is possible to identify the coding strand of the cDNA clone. Provided a sufficient length of sequence has been established to make their presence statistically likely (in practice 300–500 nucleotides), termination codons will occur in two of the three potential coding frames, ruling them out as translational open reading frames (ORFs).

Sometimes this level of analysis, that is, identification of the reading frame and deduction of the encoded protein sequence, will be all that is required. However, it is often desirable to determine the structure of the gene itself. It will have been isolated as part of a genomic clone (Chapter 4). This will contain the gene and a variable amount of flanking DNA depending upon the method of cloning used.

By correlating the sequence of the gene with the sequence of the mRNA deduced from the cDNA clone, it is possible to identify the site of poly(A) addition and the positions of introns. However, technical limitations in the cDNA cloning procedure ensure that the cDNA clone will not normally contain sequences derived from the extreme 5′ terminus of the mRNA. There are two potential methods to determine the position of the 5′ end, primer extension and S1 mapping.

3.2.1 Primer extension

In primer extension, a short oligonucleotide complementary to a sequence tract near to the 5′ end of the mRNA is annealed to the mRNA. This is then used to prime synthesis with the enzyme reverse transcriptase (*Figure 3.8*). This enzyme produces a DNA copy of the mRNA (a cDNA) which extends to the 5′ terminus of the mRNA. By correlating the length of the cDNA primer

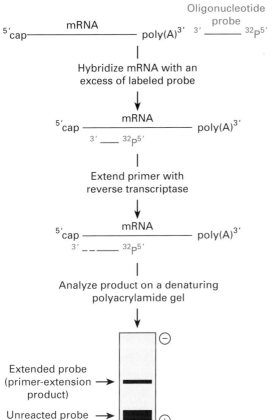

Figure 3.8: *Identification of the 5' end of an mRNA by primer extension.*

extension product with the sequence of the genomic DNA, it is possible to deduce the start site of transcription.

3.2.2 S1 mapping

The alternative technique is called S1 mapping (*Figure 3.9*). An isotopically labeled probe is generated from the genomic clone. This probe is complementary to the mRNA over a region encompassing part or all of the first exon, and extending upstream of the region where the cap site is expected to be located. The cap site may have already been putatively localized using primer extension. The probe is annealed to the mRNA and digested with an enzyme which destroys single-stranded nucleic acids but leaves double-stranded molecules intact. In early experiments an enzyme called nuclease S1 was generally used, and the name S1 mapping has been retained, even though other nucleases are now sometimes used. During digestion, the 3' proximal region of the probe, which has no complement in the mRNA molecule and which therefore remains single-stranded, is degraded. A molecule is generated which is the complement of the region between the cap site and the position at

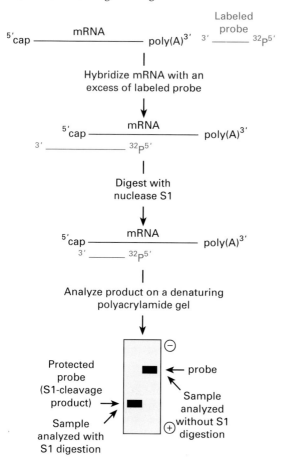

Figure 3.9: Identification of the 5' end of an mRNA by S1 mapping.

which the 5′ terminus of the probe annealed to the mRNA. Again, by determining the length of this molecule it is possible to infer the position of the cap site.

These two techniques are often used in parallel to determine the cap site. They are both subject to artefacts, but of different kinds. Therefore if both methods agree, it greatly increases confidence in the assignment of the cap site. If the hybridization is performed with an excess of probe, the amount of primer extension or nuclease-resistant product is proportional to the amount of complementary mRNA. Both techniques can therefore be used as an alternative to Northern transfer to quantitate mRNA levels.

3.3 Searching for genes using computers

Although the series of analyses described above are necessary ultimately to establish the organization and mode of expression of a particular gene, genes can be sought by less laborious means using computer programs. These search for regions of a DNA sequence which look like coding regions flanked by the signals that direct gene transcription and processing of their RNA products

(*Figure 3.10*). This procedure is of greatest value when large regions of genomic DNA for which cDNA clones are not available are being analyzed. It will thus be an essential part of the Human Genome Project.

One of the principles, that of searching for long ORFs, has already been described. Although the sequence (the Kozak sequence) that precedes the ATG initiation codon is only weakly conserved between different genes it can provide a helpful guide as to the start site of translation. The start site of transcription often lies downstream of a TATA box, and it is also possible to search for the consensus binding sites of other transcription factors. Binding

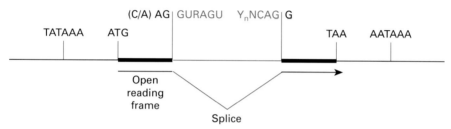

Figure 3.10: *Searching for genes using computers. Structure of a gene shown as consensus sequences.*

sites for Sp1 (Chapter 1) are particularly common, both immediately upstream of 'housekeeping' genes and within their 5′ proximal coding regions. A consensus Sp1 site contains a CpG dinucleotide, a dinucleotide which is greatly under-represented in other parts of the mammalian genome. Hence the 5′ end of genes which have such a structure are sometimes called CpG islands [5]. These regions can be identified, even in the absence of DNA sequence information, because they are susceptible to cleavage by rare-cutting restriction enzymes such as *Not*I (Chapter 2). Their presence at the 5′ end of a long ORF, or series of ORFs, is a strong indicator that a gene is present.

RNA processing signals can also be used to help establish the likely gene organization. The sequence that signals polyadenylation is well conserved, and splice junctions have a consensus sequence comprising an almost perfectly conserved core (5′ GU−−−−AG 3′), with moderately well conserved flanking sequences (*Figure 3.10*).

Identification of genes by analysis of DNA sequences is an evolving science. The structures of new genes are constantly adding to the rules, and the sophistication of the search programs is increasing. However, at present it is used largely as a guide to further action, that is, to the application of the precise mapping techniques described above. Such deduced gene structures can, however, be used to determine whether a similar gene has ever been observed before. This is done by computer searches of protein or DNA databases. These databanks contain many thousands of gene sequences. At the time of writing there are two main databases, Genbank and the EMBL. Although some scientists initially submit their gene sequence data to one rather than other, the two organizations that compile the databases then exchange data.

At the time of writing there are 71 280 sequences in the Genbank database and 72 481 sequences in the EMBL database. These numbers will grow by tens of thousands over the coming years. Such are the sizes of these two banks that there is a reasonable likelihood of finding a homolog to any gene sequence used in the search. Thus, in a screen of 2000 cDNA clones derived from brain tissue, 50% showed some homology to a known protein [6]. Even if the whole of the gene has no homolog in the database, very often a part of the gene will be homologous to another gene. This happens because genes are generally built up of modules, for example, an actin-binding domain linked to a calcium-binding domain, or a DNA binding domain linked to a potential transcriptional activation domain. Identifying a module within a gene can be an invaluable clue to its likely function.

3.4 The polymerase chain reaction

The polymerase chain reaction (PCR) has become one of the most valuable techniques in molecular biology and underlies many of the most exciting advances made in recent years [7]. Relying only upon a knowledge of the nucleic acid sequence of a gene and its functional organization, it allows the synthesis of microgram amounts of specific nucleic acid sequences from any part of the genome. It can, for example, be used as a powerful diagnostic test for the presence of mutations within the human genome, or to introduce specific mutations into a cloned gene.

Two oligonucleotide primers are synthesized which derive from opposite strands of the template DNA to be amplified and which have their 3′ termini facing each other (*Figure 3.11*). The target DNA is denatured in the presence of a large excess of the two primers and then returned to a temperature which will allow the primers to anneal to the DNA. A heat-stable DNA polymerase (most frequently that isolated from the thermophilic bacterium *T. aquaticus* and therefore called *Taq* polymerase), and all four nucleoside triphosphates, are included in the reaction and the sample is placed at a temperature optimal for elongation by the enzyme. This produces two copies of the sequence lying between the primers. If the reaction has occurred for a sufficient time, the sequences will extend beyond the position at which the oppositely oriented primer annealed.

The above three steps of melting, annealing and extension are repeated. This produces another four copies of the sequence bracketed by the oligonucleotide primers. Again, two of these will have indeterminate 3′ termini but two will now have termini dictated by the 5′ terminus of the other primer. In the subsequent cycles of reactions there will be further amplification such that, within a few cycles, the predominant product is that defined by the 5′ termini of the starting primers. Originally, PCR was performed manually, by laboriously transferring the test tubes between water baths at the required temperatures. Now, however, commercially available machines are used in which the tubes sit within a metal heating block that is programmed to cycle rapidly between the required temperatures.

In principle, n cycles of PCR amplify the target 2^n-fold. *Taq* polymerase is very thermostable, and 40 or 50 cycles of amplification can be performed if

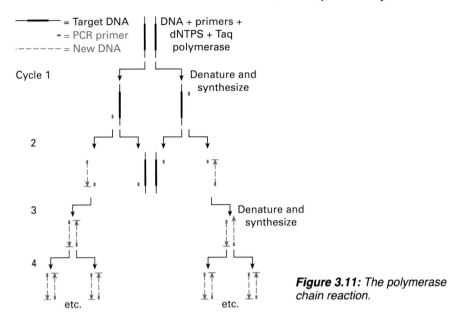

Figure 3.11: *The polymerase chain reaction.*

necessary. Hence as little as a single molecule of target can generate a detectable amount of PCR product. It is therefore an analytical technique of immense power. In fact one of the major problems to be overcome when using PCR at this level of detection is over-sensitivity. False positives are all too readily generated by trace amounts of the target DNA in the laboratory environment.

Again, because the technique is so sensitive, there is also a risk of generating false positives from the target added to the tube. This occurs when one of the primers cross-hybridizes to sequences within the target DNA that bear some degree of homology to the intended target sequence. This can, in the worst case, lead to a whole welter of bands on the analytical gel, only one of which is the authentic reaction product. This problem can usually be resolved by altering the PCR conditions to minimize cross-hybridization. Since the ionic strength is one of the factors determining T_m, this is most conveniently done by decreasing the magnesium concentration (*Figure 3.12*). Alternatively, the annealing temperature can be increased.

Another factor which will help determine the precise PCR conditions is the length of the product to be amplified, that is, the distance between the two PCR primers. It is possible to produce extremely long products – one case has been reported of two primers being separated by over 10 kb [8]. However, this is an exception. Although it is possible to produce fragments of a few thousand bases in length, the average product length is in the region of 100–500 bp. Given the empirical nature of the optimization, and the fact that artefacts can occur, how can one be confident that the band observed on a gel is the correct product? One obvious criterion is that it should be the expected size. Another is that when the magnesium concentration is decreased (that is, at increased stringency) the proportion of the authentic band in the mixture of products should

Increasing concentration of Mg^{2+}

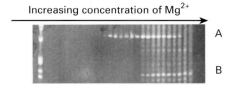

A

B

Figure 3.12: *PCR reaction showing the effect of increasing concentrations of magnesium in the reaction buffer. A and B are two different genes amplified together and showing considerable variation in the amount of product produced depending on magnesium concentration. The PCR products were separated on an acrylamide gel and visualized under UV light after staining with ethidium bromide.*

be selectively increased. If, however, it is essential that a product be unambiguously identified, then the sequence of the PCR product must be determined. One method for achieving this is shown in *Figure 3.13*.

Determining the sequence of PCR products, isolated as individual chains by molecular cloning, has revealed that there is an error rate in their synthesis. This can be as high as one misincorporated base per thousand synthesized [9]. This is not a problem when PCR is being used analytically, since the size of the product and/or its ability to hybridize to a complementary DNA probe will not be affected. It can, however, be a problem when the aim is to clone the PCR product for some specific use. The only absolutely safe procedure is to determine the entire DNA sequence of the PCR product to check for errors. Since errors are introduced essentially randomly, it will normally be possible to obtain a clone containing the authentic sequence.

3.5 Forensic and medical applications of PCR

3.5.1 Microsatellites

In Chapter 2 two classes of DNA markers detecting polymorphic variation were described, restriction fragment length polymorphisms (RFLPs) and variable number tandem repeats (VNTRs). There is a third class of highly variable markers, termed microsatellites, that are best detected using PCR (*Figure 3.14*). They are composed of di-, tri- and tetra-nucleotide repeats, for example $(CA)_n$ or $(CCA)_n$, where n can vary from 10 to more than 30. A typical example of a microsatellite on the human Y chromosome is shown in *Figure 3.14*. In this particular individual there are 17 repeats of a CA sequence. Microsatellites are simple to use and, compared to Southern blotting, they reduce the time taken to score a polymorphism from 1–2 weeks to less than 2 days. Microsatellite markers are often more informative than conventional RFLPs because they have many possible alleles. Also, they appear to be

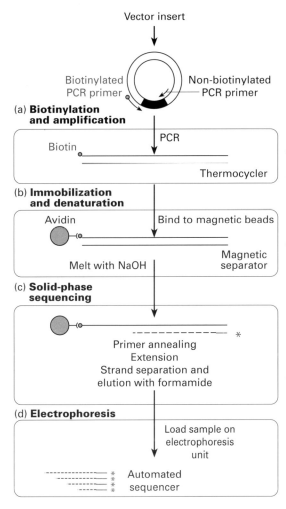

Figure 3.13: PCR sequence analysis using biotinylated oligonucleotides. This helps produce PCR products which are suitable as templates for sequencing. One of the two PCR primers is modified by incorporating a biotinylated deoxynucleotide as the final base during synthesis of the primer. After PCR with the modified primers, the product is isolated by adding streptavidin-coated magnetic beads to the reaction mix. Biotin has an extremely strong affinity for streptavidin, and the biotinylated product will bind to the magnetic beads. The beads are then purified away from the reaction using a magnet. This is a double-stranded product; single-stranded template for sequencing is obtained by mild treatment with alkali. The strand bound to the beads is washed and used as template for chain termination sequencing (Figure 3.14).

distributed randomly across the genome, unlike conventional VNTR markers which tend to be concentrated near the ends of chromosomes.

Markers defined using PCR require only minimal amounts of DNA and therefore microsatellites are increasingly being used in forensic medicine, paternity testing and medical diagnostics. An important use of microsatellites is as an alternative to Southern transfer in forensic analysis of DNA. Its great advantage is that tiny amounts of tissue, literally a blood spot, will give enough DNA to yield a result. For purposes of genetic counselling, it is possible to identify cystic fibrosis gene carriers using a few cells obtained from a mouth-wash. This is a highly acceptable form of testing large numbers of individuals as it is non-invasive, compared to taking blood. In the next few years, these markers are likely to become the diagnostic tools of choice and PCR may therefore largely replace Southern blot hybridization as a means of genetic diagnosis.

3.5.2 Diagnosing infectious diseases by PCR

Another application of PCR for diagnostic purposes is in the identification of DNA brought into the patient's tissues by an infectious agent. This technique offers several advantages over the more commonly used serological and biological procedures. Because PCR is so sensitive it allows an earlier diagnosis to be made. Also, when direct identification of a viable infectious agent is not possible, detection of that agent's DNA is considered good evidence of its viability and potential to self-replicate within the host. In contrast, serology

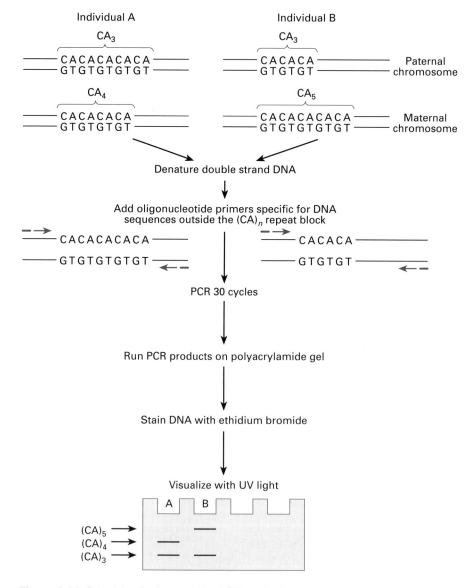

Figure 3.14: Principle of microsatellite PCR analysis.

identifies antigens that may have been released into the bloodstream. These may be no longer associated with viable viral or bacterial particles. PCR analysis can also be of particular value with immunosupressed patients, where detection of specific antibodies directed against the infecting agent is difficult. It is similarly useful in analyzing fetal blood, as not all maternal antibodies can cross the placental barrier and the fetus itself synthesizes extremely small amounts of antibodies.

In patients infected with hepatitis B virus (HBV), clinical evaluation, paralleled by ordinary serological data, is sufficient for a diagnosis to be made. The situation is complicated in the case of a chronic carrier. Direct identification of the HBV DNA in the patient's blood proves that the infectious state is still present. Identification of the 'e' antigen (a protein present in the core of the viral particle), although of the same significance, is not as sensitive. Proof that the viral infection is still active is also of prognostic importance, because it is associated with development and persistence of chronic hepatic damage (chronic hepatic cirrhosis or hepatocarcinoma).

In the case of congenital toxoplasmosis, serological analysis of fetal blood is virtually ineffective and here PCR is of great value. Fetal infection causes a number of pathological situations depending on the period of pregnancy during which infection has occurred. These range from oculopathy to gross malformations in the central nervous system with resultant irreversible brain damage. The small size of the fetal blood specimen, and the low concentration of the microbial DNA, make the sensitivity of the PCR reaction a great advantage. Pharmacological therapy can give good results only if an early diagnosis, prior to irreversible brain damage or malformations, can be made. Again, serology is of no use in immunodepressed AIDS patients, where toxoplasmosis is among the most common opportunistic infections.

Other infectious agents for which PCR analysis is very valuable are cytomegalovirus (CMV), chlamydiae, and herpes virus (HV). These viruses affect the infant at the time of birth. The conventional analysis methods are, in the case of HV or chlamydial infections, immunohistochemical characterization of uterine cervix cells and, for CMV, isolation from cultured neonatal urinary sediment cells. Such approaches are time-consuming, while PCR analysis allows rapid identification even if minimal amounts of the virus are present.

References

1. Maxam, A.M. and Gilbert, W. (1977) *Proc. Natl. Acad. Sci. USA*, **74**, 560.
2. Sanger, F., Coulson, A.R., Barrell, B.G., Smith, A.J.H. and Roe, B.A. (1980) *J. Mol. Biol.*, **143**, 161.
3. Martin-Gollardo, A., McCombie, W.R., Gocayne, J.D., *et al.* (1992) *Nature Genetics*, **1**, 34.
4. Oliver, S.G., van der Aart, G.M., Agastoni-Carbone, M.L., *et al.* (1992) *Nature*, **357**, 38.
5. Bird, A.P. (1986) *Nature*, **321**, 209.
6. Adams, M.D., Dubnik, M., Kerlavage, A.R., *et al.* (1992) *Nature*, **355**, 632.
7. Saiki, R.K., Scharf, S.J. Faloona, F., *et al.* (1985) *Science*, **230**, 1350.
8. Jeffreys, A.J., Wilson, V., Neumann, R. and Keye, J. (1988) *Nucl. Acids Res.*, **16**, 10953.
9. Eckert, K.A. and Kunkel, T.A. (1990) *Nucl. Acids Res.*, **18**, 3739.
10. Dunning, A.M. (1988) *Nucl. Acids Res.*, **16**, 10393.

Further reading

Ballantyne, J., Sensabaugh, G. and Witkowski, J.A. (1989) *DNA Technology and Forensic Science*, Banbury Report number 32. Cold Spring Harbor Laboratory, New York.

Gill, P., Jeffreys, A.J. and Werrett, D.J. (1985) Forensic applications of DNA fingerprints. *Nature*, **318**, 577.

Innis, M.A., Yeland, D.H., Sninsky, J.J. and White, T.J. (1990)*PCR Protocols: A Guide to Methods and Applications*. Academic Press, San Diego.

Kogan S.C., Doherty, M. and Gitschier, J. (1987) An improved method for prenatal diagnosis of genetic diseases by analysis of amplified DNA sequences. *New Engl. J. Med.*, **317**; 985.

Paabo, S., Higuchi, R.G. and Wilson, A.C. (1989) Ancient DNA and the polymerase chain reaction. *J. Biol. Chem.*, **264**, 9709.

Reiss, J. and Cooper, D.N. (1990) Application of the polymerase chain reaction to the diagnosis of human genetic disease. *Hum. Genet.*, **85**, 1–8.

4

CLONING IN *E. COLI.* I: VECTORS AND GENOMIC LIBRARIES

4.1 Cloning and cloning vectors

In gene cloning we introduce into a host cell a vector, which is a DNA molecule with the capacity to replicate in parallel with the endogenous genome. Here we describe cloning in the bacterial host *E. coli*, which is the general workhorse of genetic engineering. Cloning involves four basic steps:

(a) the vector DNA is cleaved with one or more restriction enzymes;
(b) the DNA to be cloned, the target or insert, is joined to the vector, generating a recombinant molecule;
(c) the recombinant DNA molecules are introduced into the host bacterial cell, which is said be transformed by the introduction of the vector molecule;
(d) if a mixture of recombinant molecules is introduced into a host cell population at low efficiency, so that most cells receive only one DNA molecule, then each colony of cells that grows up on an agar plate will contain a different recombinant DNA molecule. Each colony can be picked separately, or cloned, and the particular recombinant molecule that it contains can be isolated, identified and amplified.

Cloning constitutes a one-step purification of such enormous power that single genes, comprising only one part in one million of total human genomic DNA, can readily be isolated. While PCR offers the possibility of amplifying specific segments of DNA in the test tube, there are many purposes for which gene cloning remains essential. These include the initial isolation of a gene, its propagation in the laboratory and the production of its protein product. The two methods, PCR and gene cloning, should be viewed as complementary techniques. Molecular cloning of a PCR product is now one of the most commonly used routes to isolating and manipulating genes.

A cloning vector is necessary primarily as a carrier of the cloned gene. Without a vector, a DNA molecule introduced into a cell would be diluted out by cell division and eventually lost. There are a number of features common to all cloning vectors. First, the vector must be able to replicate in the host cell. Hence it must have a replication origin, a sequence element that is recognized by the host cell's replication machinery and that directs amplification of the vector. Secondly, transformation is a somewhat inefficient process. Only a minority of cells in the population take up and retain the exogenous DNA.

Therefore a second requirement is that the vector must contain a selectable marker, so that the subpopulation of bacteria containing it can be isolated.

Some vectors are based upon extrachromosomal circular DNA molecules called plasmids. Plasmid cloning vectors derive from naturally occurring drug resistance plasmids. An antibiotic resistance gene normally encodes an enzyme which modifies or breaks down the antibiotic — for example, ampicillin resistance is conferred by a β-lactamase which degrades the drug. Clones of transformant cells containing a drug resistance plasmid are selected by their ability to grow in the presence of a concentration of the antibiotic which prevents the growth of, or is toxic to, non-transformed cells.

There are also vectors based upon bacterial DNA viruses, such as bacteriophage λ. When a λ phage particle infects a cell, circular areas of dead bacteria called plaques are formed in the 'lawn' of healthy cells. Each plaque contains the viral progeny of a single infectious particle, and the virus can be picked from it and amplified by growth in a fresh host cell.

Aside from these two absolute essentials, that is, a replication origin and a method of selection, most vectors in common use contain a short DNA sequence containing many closely spaced restriction enzyme cleavage sites. Such a multi-cloning site (MCS) or polylinker is constructed by chemical synthesis. An MCS increases the number of potential cloning strategies available, by extending the range of enzymes that can be used to generate a restriction fragment suitable for cloning. By combining them within an MCS the sites are made contiguous, so that any two sites within it can be cleaved simultaneously without excising vector sequences.

Thus there are three basic steps in any cloning procedure, constructing the recombinant DNA molecule, introducing it into the bacterium, and selecting a transformed cell containing the desired recombinant. These steps are performed most simply using plasmid vectors. Plasmids are very frequently used for isolating subregions (or subclones) from segments of DNA that have already been purified either by gene cloning or by PCR. Subcloning into a plasmid vector is an essential part of the technology of genetic engineering and it also illustrates the fundamental principles of all cloning procedures.

4.2 Plasmid vectors and their use in manipulating DNA

4.2.1 The pUC series of vectors

In the case of plasmid vectors, such as pUC19 [1], all the features described above are contained within a remarkably small piece of DNA, only 2.8 kb in size (*Figure 4.1*). The pUC series of vectors also incorporate a DNA sequence that allows rapid screening for the presence of an insert. The MCS lies within a short region of DNA sequence derived from the 5′ end of the *lac* operon of *E. coli* (see *Figure 1.7*). It contains the promoter and a small segment of the β-galactosidase (β-gal) gene. This encodes a sequence known as the α peptide (*Figure 4.3*). The α peptide will combine with a mutant β-gal protein which lacks the normal N-terminus, to generate a functional enzyme molecule. When an *E. coli* cell containing such a mutant β-gal gene within its genome is transformed with a pUC vector, enzymatically active β-gal accumulates. Such

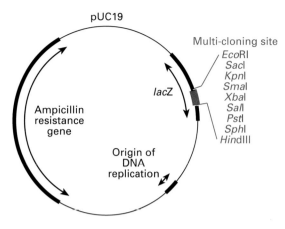

Figure 4.1: *The pUC19 vector. The other members of the pUC series of vectors differ from each other only in the nature of the restriction sites present within the MCS. The region marked* lacZ *is a short piece of DNA from the 5' end of the* lac *operon (see Figure 4.2). This drawing is not to scale.*

transformed clones can be recognized by including in the agar a β-gal substrate which produces a blue precipitate when hydrolyzed. Normally, any fragment of DNA inserted into the MCS will at some point along its length happen to contain a termination codon in the same reading frame as the α peptide. This will prevent synthesis of the α peptide. Hence bacterial colonies containing recombinant DNA molecules are colorless and can be easily identified from the 'background' of blue, non-recombinant clones.

4.2.2 DNA subcloning

DNA subcloning using a plasmid vector involves the following steps (*Figure 4.2*):

(a) the target to be cloned is cleaved from the source DNA molecule;
(b) the vector molecule is cleaved in its MCS;
(c) the target and vector molecules are joined;
(d) *E. coli* cells are transformed using the recombinant vector.

For the first step, cleavage of the target sequence from the source, we use either a single restriction enzyme or a pair of enzymes. If there are no restriction enzyme sites at the required positions within the starting molecule, then PCR is often used to generate a product with the desired termini. By incorporating restriction sites into the PCR primers (at the 5' end, where pairing with the template is not critical), total control can be obtained over the nature of the final product. The target is usually purified on an agarose or polyacrylamide gel to remove unwanted molecules which would interfere with cloning.

Since plasmids are circular, the second step, cleaving the vector within the MCS, is often referred to as linearization. This is done with an enzyme, or with two enzymes, that generate ends compatible with the termini on the target (*Figure 4.2*). In the simplest case the ends will be blunt. Any blunt-ended molecule will join to any other blunt-ended molecule. However, the efficiency of joining is lower for blunt than for staggered ends, because the annealing of cohesive termini greatly favors their joining. If possible, therefore, enzymes which generate sticky ends are used.

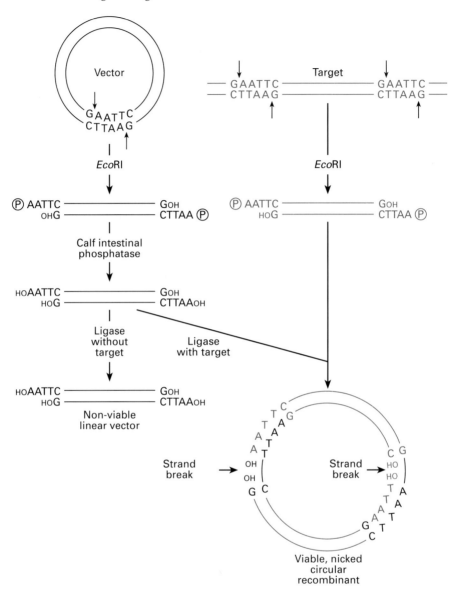

Figure 4.2: *A typical subcloning experiment. Redrawn from Williams and Patient (1989) with permission from Oxford University Press.*

These need not necessarily be the same enzyme used to generate the target, but the enzymes used to cleave the vector and the target must produce compatible overhanging termini. *Sau*3A recognizes the sequence GATC, and *Bam*HI recognizes the sequence GGATCC, but both enzymes generate a 5′ overhanging terminus with the same sequence, GATC. Hence a *Sau*3A fragment can be cloned into a *Bam*HI site. If two different sticky-end enzymes are used to generate the target, then the fragment will have a fixed orientation with respect to the vector molecule. This is sometimes referrred to as a 'forced' cloning. If a single enzyme is used then the fragment may come to lie in either orientation.

The third step, joining the target and vector molecules, uses the enzyme T4 DNA ligase. The desired reaction is a two step ligation, in which a molecule of target joins first to one end, and then to the other end, of a vector molecule. This will generate a circular molecule containing the target DNA. Linear DNA has a transformation efficiency several orders of magnitude lower than circular DNA. Hence recombinant molecules are cloned much more frequently than linear vector molecules. However, there are a number of competing ligation reactions, such as end-to-end joining of vector or target molecules and recircularization of the vector without a target molecule.

The competing reactions can be controlled to an extent by varying the ratios of enzyme and target, but recircularization of the linearized vector is a very frequent event (because the ends are held in such close relative proximity) and cannot be completely eliminated in this way. Vector recircularization generates a background of non-transformant clones that will greatly exceed the number of clones containing recombinant DNA molecules. A number of tricks are available to minimize the problem.

One is to accept the high background but to identify recombinants by cloning into an MCS, such as that of the pUC vectors, where β-gal screening selection can be performed (*Figure 4.3*). Alternatively, if two restriction enzymes which generate different cohesive termini are used to cleave the vector for cloning (that is, if forced cloning is performed) then the background will be greatly reduced, because the vector cannot recircularize. If neither of

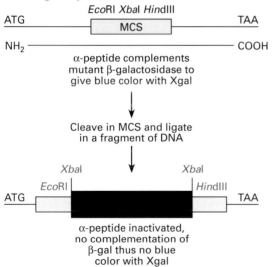

Figure 4.3: *The principle of* lacZ *selection. An MCS of this form is present in many plasmid and phage vectors. As well as allowing application of a simple blue–white test for the presence of an insert, the* lac *fragment provides a strong bacterial promoter and Shine–Dalgarno signal. Thus genes inserted into such an MCS, in the correct orientation and in frame with the β-gal ATG initiation codon, are expressed in* E. coli *at a high level. If the resulting fusion protein is stable, it can be isolated and analyzed further or utilized. Redrawn from Williams and Patient (1989) with permission from Oxford University Press.*

these options is practicable, then the background can be reduced by treating the vector with a phosphatase prior to ligation (*Figure 4.2*). Phosphatase removes the 5′ terminal phosphate groups from the termini. T4 DNA ligase requires a 5′ phosphate in order to join two molecules, so vector molecules cannot recircularize. However, target molecules can ligate to vector molecules to form a circular molecule. Those vector molecules which recircularize by joining with a target molecule are able to transform *E. coli*. Although only one covalent bond is formed at each junction of vector with target (using the phosphate donated by the target), this is sufficient to keep the vector and target molecules attached and the gaps are repaired once the DNA is introduced into *E. coli*.

The fourth step, transformation, can be done in a number of different ways. Most commonly, bacterial membranes are rendered temporarily permeable to DNA by exposure to divalent cations or by a brief electric shock (electroporation). The bacteria are then placed under conditions selective for the growth of transformant cells, normally by plating on an agar plate in the presence of an antibiotic. If the target is pure, and the procedure used to select against vector recircularization successful, most or all of the bacterial clones will contain the desired sequence. However, this is normally checked by growing small bacterial cultures (mini-preparations) derived from individual colonies. The cells are lysed, plasmid DNA is extracted and a restriction enzyme cleavage map determined to verify that it has the expected structure. Only then is a large scale plasmid preparation made by extraction with organic solvents, by centrifugation or by column chromatography.

4.2.3 Site-directed mutagenesis

Once a gene has been isolated by cloning, it is sometimes necessary to modify its sequence in some way, perhaps to change the biological properties of the encoded protein or to optimize its expression in a host organism. This is achieved by site-directed mutagenesis. A method in common use is shown in *Figure 4.4*. It relies on a series of PCR steps, one of which uses primers that incorporate the changes to be made [2]. These mismatches will partially destabilize the annealed product, but provided there is a sufficient region of perfectly homologous sequence flanking the mismatch, a hybrid will be formed which can be recovered by cloning in *E. coli*.

4.2.4 Multi-purpose plasmid vectors

Vector choice is very much a case of selecting 'horses for courses'. Simple plasmid vectors, such as those in the pUC series, still find many applications, but a number of more complex vectors which allow additional manipulations of the cloned target, are now in more common use.

Bacteriophage RNA polymerases, such as that derived from phage T7, will initiate transcription only at a promoter sequence found upstream of their own genes. In order to allow the production of a single-stranded RNA copy of the insert, many vectors incorporate chemically synthesized copies of specific bacteriophage promoters. There are normally two promoters from different

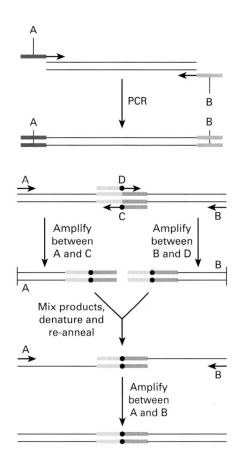

Figure 4.4: *Site-directed mutagenesis. The region encompassing the site to be mutagenized is first isolated by PCR using primers that bear restriction sites at their 5' termini. A second set of overlapping primers are then synthesized which are complementary to the position where the mutation is to be inserted (the changed sequence is represented by a filled circle). This gives two products, a 5'-derived and a 3'-derived fragment, both containing the mutation. These are then mixed, denatured and re-annealed to give a mainly single-stranded copy of the mutant version of the gene. This is finally converted into a double-stranded form by PCR using the original two oligonucleotides as primers. This product is then cloned into a suitable vector and transformed into* E. coli.

phage flanking the MCS, for example the T3 phage promoter on one side and the T7 promoter on the other (*Figure 4.5*). By cloning a gene into the MCS and copying with the appropriate polymerase, RNA transcripts which have the same sequence as the gene ('sense' RNA) or the complementary sequence ('anti-sense' RNA) are generated. Sense and anti-sense RNAs have many uses. For example, sense RNA can be used to direct synthesis of the cognate protein, using an *in vitro* translation system derived from rabbit reticulocytes. Anti-sense RNA can be used as a hybridization probe to detect transcripts of the gene by Northern blotting.

It is sometimes useful to be able to generate single-stranded DNA copies of the insert in a recombinant molecule. This may be to provide a template for DNA sequence analysis, or so that a single-stranded isotopically labeled hybridization probe can be prepared. Many vectors incorporate the replication origin of the single-stranded DNA bacteriophage M13. When bacteria containing such a vector are infected with 'helper' M13 phage, which encode the proteins needed for viral replication, single-stranded DNA copies of the plasmid vector are produced and packaged into pseudo-viral particles. Single-stranded DNA can be isolated from these particles by lysing them and purifying the DNA away from the phage proteins.

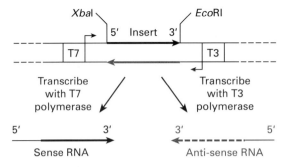

Figure 4.5: *Sense and anti-sense RNA production using bacteriophage polymerases. The coding strand of the insert is shown as a solid black arrow. When the insert DNA is copied with T7 polymerase the transcript will have the sequence of the coding strand of the gene from which it derives, i.e. if translated it would yield the cognate protein. Hence it is called 'sense' RNA. When copied with T3 polymerase the transcript will have the sequence of the non-coding strand of the gene and is therefore called 'anti-sense' RNA.*

4.3 Phage vectors

4.3.1 M13 vectors

When the only specialized requirement is that single-stranded DNA be easily prepared, derivatives of the bacteriophage M13 itself are often used as cloning

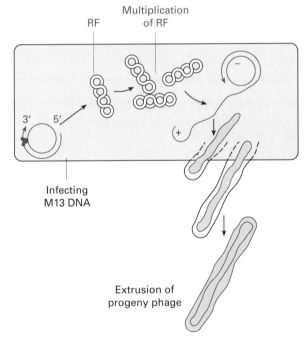

Figure 4.6: *Production of single-stranded DNA. Inflecting single-stranded DNA is converted into double-stranded replicative form (RF) DNA. After replication one strand (the + strand) is packaged into new progeny viral particles. Redrawn from Williams and Patient (1989) with permission from Oxford University Press.*

vectors (*Figure 4.6*). This removes the requirement for co-infection with a helper phage, but transformation efficiencies for M13 are relatively low, and large inserts are subject to frequent re-arrangements during propagation. The gene of interest will usually therefore have been isolated by cloning into a vector that accepts large inserts readily. Fragments are then subcloned into M13 and single-stranded DNA prepared for sequence analysis.

4.3.2 In vitro *packaging of bacteriophage lambda DNA*

The double-stranded DNA virus, bacteriophage λ, is extremely well character-ized genetically and biochemically, allowing elegant methods to be used to simplify the task of gene cloning. One of its most useful features is the ability to assemble phage particles *in vitro*.

The λ genome is a linear, double-stranded DNA molecule 45 kb in length. It replicates to form a long concatemer, containing many viral genomes attached end-to-end. Because the termini of the viral genome have staggered ends, they are mutually cohesive. Hence they are called *cos* sites. During packaging of the viral DNA into phage heads, the *cos* sites act as targets for cleavage of the multimeric DNA, to yield separate viral genomes. An extract prepared from infected cells will incorporate added λ DNA into phage coat proteins to yield viral particles. This is called *in vitro* packaging [3]. The only viral sequences required for packaging are the *cos* sites, so that λ vectors containing target sequences (recombinant phage) can be packaged.

Using such extracts, a very high proportion of input DNA is packaged, and the introduction of the resultant phage into cells by infection (transfection) is also a very efficient process. When a complex DNA, such as the entire human genome, is the target, many hundreds of thousands of DNA segments must be cloned to have a reasonable chance of isolating a specific gene. A mixture of clones derived from a complex target is termed a clone bank, or sometimes a gene library. The high cloning efficiencies which can be obtained by *in vitro* packaging minimize the amount of DNA needed to generate such a library.

There are two kinds of λ vector which differ in the size of insert they accept, λ replacement vectors and λ insertion vectors. These are described below.

4.3.3 Lambda replacement vectors

Lambda replacement vectors contain restriction sites flanking a region of the genome which is dispensable for phage propagation in suitable bacterial host strains (*Figure 4.7*). Cloning into a replacement vector is performed in a somewhat similar manner to plasmid cloning. The phage DNA is circularized by ligation of the *cos* sites. The central, non-essential fragment is cleaved out and the linked flanking fragments (arms) are purified by centrifugation or electrophoresis. Ligation is performed at a ratio of arms to target that favors the formation of very long concatemers, in which vector and target molecules are interspersed. Because *in vitro* packaging requires the two *cos* sites to be a minimum of 38 kb apart, there is an automatic selection for those molecules where a target fragment is flanked at each side by a vector molecule.

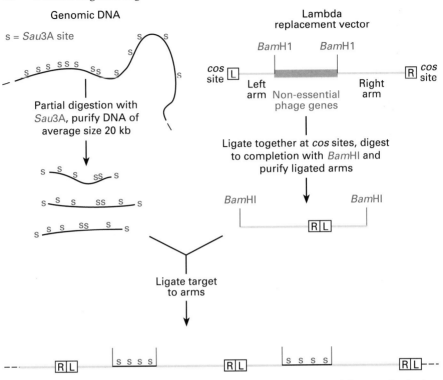

Figure 4.7: *Gene cloning in a λ replacement vector. For successful genomic cloning the target DNA must be isolated in as large a form as possible. Partial cleavage with Sau3A is performed by limiting either the time of digestion or amount of enzyme. In order to obtain efficient ligation and packaging, the ratio of arms to target, and their absolute concentration, must be carefully controlled.*

In most replacement vectors (e.g. EMBL4) [4] the internal region that is replaced by the target contains a gene that renders the phage inviable in an appropriate *E. coli* host. Hence it is possible to select against parental phage, that is, those vector molecules that do not contain target DNA, by using such a host for infection. Recombinant phage in which the internal region is replaced by target DNA will of course be viable. Because they are able to accept large DNA fragments, replacement vectors are most often used for cloning genomic DNA fragments from higher eukaryotes.

4.3.4 Lambda insertion vectors

When cloning into an insertion vector, the phage DNA is cleaved with a restriction enzyme that cuts it only once, and the target is inserted into this site (*Figure 4.8*). Because no phage DNA is removed, and the maximal separation between *cos* sites that permits DNA to be packaged is 52 kb, there is a much lower limit on the size of target which can be cloned in insertion vectors. Hence they are commonly used for cloning cDNA copies of eukaryotic mRNA sequences. These have an average size of 1.5–2 kb and do not generally exceed 5 kb in length. In the commonly used vector λgt10 [5] the *Eco*RI cloning site is

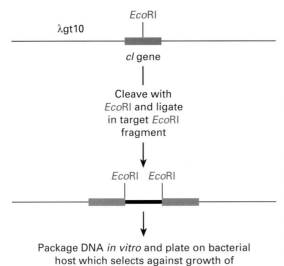

Package DNA *in vitro* and plate on bacterial host which selects against growth of non-recombinant phage (because they have an intact *cl* gene)

Figure 4.8: *Gene cloning in a* λ *insertion vector. Phage containing an intact* cl *gene will not grow on the* E. coli *strain. Thus by infecting such a host with a phage–target ligation mix, which always contains a mixture of recombinant (insert-containing) and non-recombinant (parental) phage, it is possible to filter out the unwanted non-recombinant phage completely.*

in a gene which is deleterious to phage replication in certain host strains. This allows selection against non-recombinant phage (*Figure 4.8*).

4.4 Cosmid and P1 vectors

Lambda replacement vectors permit cloning of DNA fragments of up to 20 kb, but sometimes it is necessary to isolate fragments larger than this. Possible reasons include:

(a) many genes in higher eukaryotes are longer than 20 kb and it is highly desirable to keep them in one piece during cloning;

(b) some of the procedures used to screen genomic clone banks yield clones which are close to, but not at, the position of the required locus (see Section 5.3). It is then necessary to 'walk' along the genome, by isolating a series of genomic clones which overlap, and therefore link, the starting clone and the required gene (*Figure 4.9*). It is obviously more convenient if individual steps in the walk be as large as possible, so the genomic clones should be as long as possible.

Cosmid vectors such as pJB8 [6] are commonly used to prepare genomic DNA libraries for both these purposes. As their name suggests, cosmids are plasmid vectors containing the *cos* sites of bacteriophage λ (*Figure 4.10*). Target DNA fragments are cloned into cosmid vectors in much the same way as into a λ replacement vector, and the DNA is packaged *in vitro* and introduced into *E. coli* by transfection. The cosmid vector contains a drug resistance gene, and recombinant clones are selected and propagated in just the same way as bacteria transformed with a plasmid vector. Again, this is a very efficient cloning method. Representative libraries containing inserts of up to 45 kb in length can readily be generated in cosmid vectors.

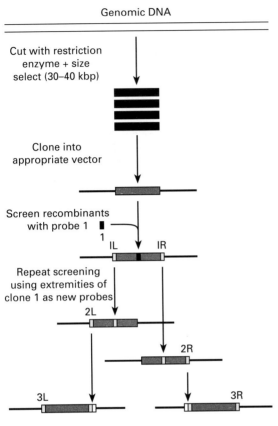

Genomic DNA

Cut with restriction
enzyme + size
select (30–40 kbp)

Clone into
appropriate vector

Screen recombinants
with probe 1

1

IL   IR

Repeat screening
using extremities of
clone 1 as new probes

2L

2R

3L   3R

Figure 4.9: *Chromosome walking. After constructing the genomic library, the screening is performed starting from a known position (probe 1). The sequences taken from the extremities of each positive clone isolated with probe 1 generate two families of probes (L=left; R=right) with which it is possible to 'walk' along a chromosome in both directions.*

Recently, a new cloning system has been developed that can accommodate even longer inserts of foreign DNA, of up to 100 kb in length [7]. This system is based on the bacteriophage P1. It is used in a similar manner to λ replacement and cosmid vectors (*Figure 4.11*).

4.5 Construction and screening of genomic libraries

The construction and screening of genomic DNA libraries are now fairly routine procedures – so routine that it is even possible to buy libraries 'off the shelf' or to have them constructed to order using one's favored DNA as the target. However, it is important to understand how this is achieved, so that the limitations inherent in using such libraries become apparent.

4.5.1 Generating target DNA for genomic cloning

The genomic DNA used as the source of the target must be carefully isolated, so that it is as large as is practically possible. Prior to cloning it must be cleaved into pieces of a size suitable for cloning in the vector of choice. The aim is to generate hundreds of thousands of different DNA fragments that form a set of overlapping cleavage products. Every gene will then be represented within a set of fragments, most of which will have different end points (*Figure 4.7*). It is very important that the cleavage of the target DNA be random, otherwise some genes would not be included within DNA fragments of a clonable size.

This kind of 'semi-random' cleavage is often achieved by partial cleavage with the enzyme *Sau*3A (*Figure 4.7*). In DNA with a normal base composition the average length of fragments generated using *Sau*3A is 256 nucleotides. It is therefore very unlikely that a required gene will not have *Sau*3A sites spaced around it in such a way as to generate a fragment within the required size range for cloning. In practice, a controlled partial digestion with *Sau*3A is performed. Fragments of a size range compatible with the vector system to be used (around 20 kb if the DNA is to be cloned in a λ vector, or around 40 kb if it is to be cloned into a cosmid vector), are purified by gradient centrifugation or gel electrophoresis (*Figure 4.7*).

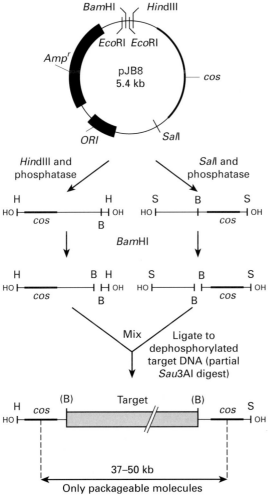

Figure 4.10: *Cosmid cloning. This complex cloning scheme [6] is designed to maximise the frequency with which a single insert fragment is flanked on each side by a vector molecule containing a* cos *site. Redrawn from Williams and Patient (1989) with permission from Oxford University Press.*

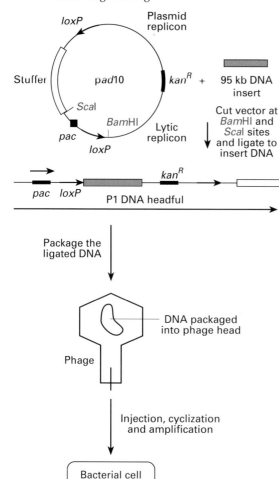

Figure 4.11: *Structure of a P1 cloning vector. The vector* contains a number of genes essential for propagation of the vector and for antibiotic selection. Up to 100 kb of target DNA can be inserted between the BamHI and ScaI sites, which generates two vector arms and disrupts the tetracycline resistance gene (tetR). *The recombinant clones are packaged* in vitro, *and used to transform bacteria to kanamycin resistance. Modified from Sternberg (1992) with permission from Elsevier.*

4.5.2 Insertion of target DNA into the cloning vector and storage of the genomic library

The target is ligated into the purified arms of a replacement vector or into a linearized cosmid vector (*Figure 4.7*). After ligation, the DNA is subjected to *in vitro* packaging and used to transfect *E. coli*. In the case of a cosmid library, the drug-resistant bacterial colonies are stored by replica plating the bacteria on to filter paper (*Figure 4.12*). This process is often termed 'lifting a library'. One copy of the library is stored frozen, so as to preserve bacterial viability. Bacteria on another, identical filter paper copy are lysed in such a way as to cause the cosmid DNA released from each colony to become denatured and bind to the filter without diffusing away. The DNA on this latter filter is subjected to *in situ* hybridization, using a radioactive probe specific for the gene of interest. Once the area on the filter containing the required clone is identified, viable bacteria are isolated from the stored copy.

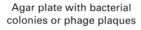

Agar plate with bacterial
colonies or phage plaques

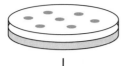

|

Transfer the bacterial colonies or phage
plaques to filter and lyse to release DNA

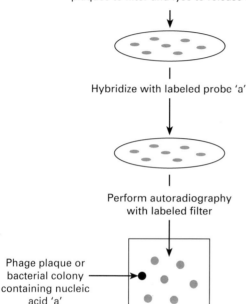

Hybridize with labeled probe 'a'

Perform autoradiography
with labeled filter

Phage plaque or
bacterial colony
containing nucleic
acid 'a'

Figure 4.12: *Screening clones by* in situ *hybridization. This technique is equally applicable to phage plaques or bacterial colonies. When screening a library, the density of plaques or phage is very much higher than shown here, with perhaps 100 000 colonies or plaques on a 15-cm diameter plate. After screening, a plug of agar is lifted out and the bacteria or phage eluted and replated at lower density. This process is repeated until a pure bacterial clone or phage population is obtained.*

In the case of λ libraries, which consist of a large number of plaques where the bacteria have been lysed rather than a collection of viable bacterial colonies, this procedure is not practicable. For long-term storage, phage particles are eluted from the plates in a mixed pool, containing representatives of every different recombinant phage on the plate. Such pooled, or amplified, libraries are very convenient since they can readily be exchanged between laboratories or sold commercially. They can then be replated onto fresh host, lifted on to filters and screened in just the same way as cosmid libraries. However, phage plaques containing different recombinant DNA molecules are almost invariably of different size, because individual cloned segments of genomic DNA may affect the extent of phage replication differently (see Section 4.5.3). Hence amplified libraries will be biased in favor of some sequences and against others.

The extent of the bias obviously increases with the number of rounds of amplification a library has undergone. It is therefore quite dangerous to use such a library to isolate a genomic clone. The sequence required may be seriously under-represented, or absent from the library. An unamplified λ library (newly packaged recombinant phage), or a filter copy of a cosmid library, are much safer, if sometimes less convenient, alternatives.

4.5.3 *Unclonable DNA and DNA re-arrangements in* E. coli

The problem of bias in genomic clone banks is part of a larger issue: how reliable is *E. coli* as a cloning host? There are sequences in mammalian DNA that are unclonable in *E. coli* and this represents a major potential problem in genomic cloning. Despite its importance, this is still an area where there are many uncertainties. Obviously, if a piece of DNA cannot be isolated by gene cloning, it is very difficult to determine the nature of the problem. This is the genetic engineer's equivalent of knowing a murder has occurred but not having access to the corpse. There are, however, general prejudices and some hard facts which can perhaps give an insight.

If a piece of DNA encodes a protein that is lethal to *E. coli*, and if this gene happens to lie downstream of sequences that can direct transcription and translation in *E. coli*, then the recombinant will never be recovered. This problem will obviously be worsened if the vector is present at high copy number, because there is normally a good proportionality between the number of copies of the gene in a cell and the amount of its protein product. Copy number is determined by plasmid-encoded sequences; for example pUC19[1] is normally present at about 500 copies per cell. There are, however, vectors that replicate to give many fewer copies per cell [8] and these are sometimes used to circumvent this problem.

Gene toxicity will sometimes lead to the absence of a particular recombinant in a gene bank, but most frequently causes the host bacterium to grow relatively slowly. This is another major reason why banks should always be screened prior to amplification if at all possible. An additional reason for avoiding amplification is the potential instability of DNA during propagation in the host. This is a particularly acute problem in the case of DNA which contains repetitive sequences. If, for example, two highly homologous genes are closely linked on the same piece of DNA in the same relative orientation, then they may undergo recombination to yield a single 'hybrid' gene. This will result in the introduction of a deletion into the cloned sequence, spanning the 3' end of one gene, the intervening DNA and the 5' end of the other gene (*Figure 4.13*). If the two genes are in the opposite relative orientation, then the host cell machinery may also catalyze recombination between the two genes, but in this case the substrate for recombination is the hairpin structure that can be formed by the two homologous sequences (*Figure 4.14*).

It is relatively rare for two highly homologous genes to be situated close together in the genome and, in order to be substrates for homologous recombination, two sequences must be identical along a considerable portion of their length. However, if a genome contains many copies of a simple repeat sequence, it is frequently subject to re-arrangement in *E. coli* because the repeats provide regions of high homology.

Recombinational deletion may occur at any time during the propagation of a particular DNA in a particular host, so that by the time a recombinant DNA is first analyzed it may already contain a deletion. It is therefore sometimes

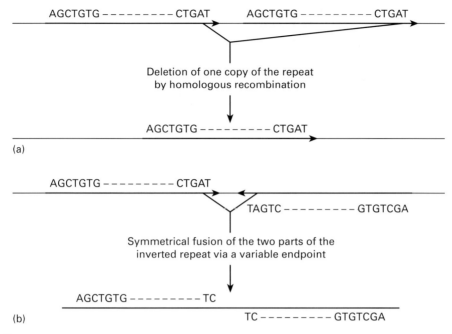

Figure 4.13: *Generation of re-arrangements by homologous recombination (a) Direct repeats. If two pieces of DNA lie in the same relative orientation, and the DNA loops back on itself such that the two homologous pieces of DNA lie side by side, then the E. coli recombination system will recognize this as a substrate for homologous recombination. The result will be to excise one copy of the repeat and the intervening DNA. (b) Inverted repeats. If two pieces of DNA lie in the opposite relative orientation and the DNA becomes supercoiled (see Chapter 1) then DNA may loop out and rearrange to form a cruciform, or snapback, structure. Such structures are also substrates for the recombination system, but the extent of the deletion may vary between clones, depending upon the site in the cruciform structure at which the rearrangement occurs.*

desirable to go back to the genomic DNA from which the gene was cloned and analyze the structure of the authentic gene by performing Southern blotting. If the cloned DNA and the genomic DNA share the same restriction map then it is reasonably safe to assume that no gross deletion has occurred.

This may all seem like a litany of potential disaster, but the risk of recombinational rearrangement can be minimized by using *E. coli* mutants with lesions in the genes catalyzing recombination. Some *E. coli* strains contain mutated versions of several different genes required for homologous recombination. If such a strain is used as a host, then the likelihood of rearrangement is greatly reduced. A price has to be paid for this in that these mutants grow relatively poorly, but this price is well worth paying if cloning artefacts can thus be avoided.

4.5.4 Chromosome jumping libraries

It will often be the case that the first step in a genomic cloning exercise yields a probe that is close to, but not overlapping, the region of interest. It is then necessary to walk along the chromosome (*Figure 4.9*). However, in any

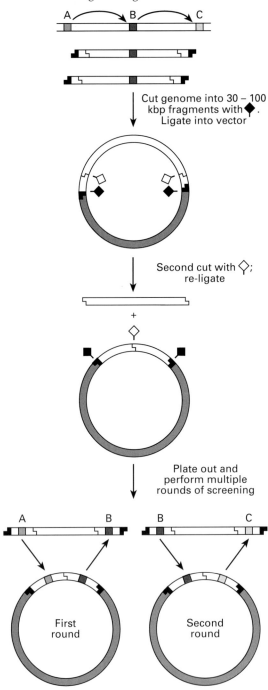

Cut genome into 30 – 100 kbp fragments with ◆. Ligate into vector

Second cut with ◇; re-ligate

Plate out and perform multiple rounds of screening

A   B

B   C

First round

Second round

Figure 4.14: *Chromosome jumping. A genomic library is constructed by cutting the genome with restriction enzyme* ◆ *. After ligating inserts to vector arms, but before inserting the DNA into host cells, the recombinant molecules are cut with a second enzyme* ◇ *that does not cut the vector. The inserts are then recircularized by ligation. In this way sequences contained between sites* ◇ *are removed. A first round of screening is carried out with the starting probe. Each round of screening is a jump over a piece of DNA that was excised during the second enzyme cleavage before cloning.*

chromosome walk, there is always the risk of encountering an unclonable region. Also, in the case of the human genome, where genetic markers are sometimes considered to be closely linked when separated by a million or more base pairs, there may be an unacceptably high number of steps in the walk. The technique of chromosome jumping [9] is designed to circumvent these problems (*Figure 4.14*). Here sequences internal to the two cloning sites are removed, prior to introduction into the bacterial host, by use of a second restriction endonuclease. This generates target DNA that falls within a reasonable size range for cloning, but in which the insert derives from two widely separated, and non-contiguous, parts of the genome. As with chromosome walking, the extremities of the cloned fragments are used as probes to move along the chromosome in a series of sequential screening and clone isolation steps. The advantage is that the steps are much larger than in orthodox chromosome walking. Rare-cutting enzymes, such as *Not*I (Chapter 2) are generally used in this procedure. Because sites for such enzymes are frequently found at the 5′ ends of genes, it will sometimes be the case that the jump will be from the 5′ end of one gene to the 5′ end of another.

References

1. Vieira, J. and Messing J. (1982) *Gene*, **19**, 259.
2. Leung, D.W., Chen, E. and Goeddel, D.V. (1989) *Technique*, **1**, 11.
3. Hohn, B. and Murray, K. (1977) *Proc. Natl. Acad. Sci. USA*, **74**, 3529.
4. Frischauf, A.M., Lehrach, H., Poutska, A. and Murray, N. (1983) *J. Mol. Biol.*, **170**, 827.
5. Huynh, T.V., Young, R.A. and Davis, R.W. (1985) in *DNA Cloning: A Practical Approach*, Vol. 1 (D.M. Glover, Ed.), IRL Press, Oxford.
6. Ish-Horowicz, D. and Burke, J.F. (1981) *Nucl. Acids Res.*, **9**, 2489.
7. Sternberg, N.L. (1992) *Trends Genet.*, **8**, 11.
8. Stoker, N.G., Fairweather, N.F. and Spratt, B.G. (1982) *Gene*, **18**, 335.
9. Lennon, G. and Lehrach, H. (1991) *Trends Genet.*, **7**, 314.

Further reading

Miller, J.H. (1972) *Experiments in Molecular Genetics*. Cold Spring Harbor Laboratory Press, New York.

Berger, S.L. and Kimmel, A.R. (1987) *A Guide to Molecular Cloning Techniques*. Methods in Enzymology, Vol. 152. Academic Press, San Diego.

Sambrook, J., Fritsch, E.F. and Maniatis, T. (1989) *Molecular Cloning, A Laboratory Manual*. Cold Spring Harbor Laboratory Press, New York.

Williams, J.G. and Patient, R.K. (1989) *Genetic Engineering*. IRL Press, Oxford.

Winnacker, E.L. (1987) *From Genes to Clones: Introduction to Gene Technology*, VCH, Weinheim.

5
CLONING IN E. COLI. II: ISOLATING GENES

5.1 Construction of cDNA libraries

If total human genomic DNA is digested to completion with a restriction enzyme that cuts it into pieces averaging 40 kb in size, over 100 000 different fragments are generated. Cloning would allow each of these to be isolated separately, but screening them individually for a specific gene would be a gargantuan task. Also, eukaryotic genes are generally interrupted by introns, and bacteria cannot splice eukaryotic gene transcripts. This precludes a very powerful screening method based upon gene expression in bacteria. Genomic cloning is not, therefore, widely used as the initial method of identifying and isolating eukaryotic genes. They are usually isolated by cloning complementary DNA copies (cDNA clones) of their cognate mRNA sequences. The cDNA clone is then used as a hybridization probe to screen a genomic library for the required gene.

5.1.1 Purification of mRNA for cDNA cloning

In order to minimize the numbers of clones which must be prepared and screened, mRNA is normally purified from a type of cell in which the required sequence is maximally abundant. This flexibility of choice constitutes one of the great advantages of cDNA cloning over genomic cloning. In some specialized tissues one, or just a few, mRNA sequences will be highly abundant. In a reticulocyte, for example, the α and β-globin mRNAs constitute the vast bulk of the whole mRNA population.

As with the preparation of genomic libraries, it is very important that the mRNA be isolated in as intact a form as possible. The great enemies here are ribonucleases. These are released when cells are lysed with detergents, or they can be introduced through contamination during mRNA purification. There are, however, potent RNAase inhibitors and well-established practices for circumventing both these problems. The mRNA is purified away from proteins and lipids using organic solvents or by centrifugation in the presence of powerful denaturants. It constitutes only about 1–2% of total cellular RNA, so it is often purified further, away from rRNA and other structural RNAs, by affinity chromatography on short tracts of poly(dT) homopolymer bound to a cellulose matrix (oligo–dT cellulose). The mRNA anneals to oligo–dT cellulose at high salt concentration by virtue of its poly(A) tail, while rRNA and tRNA pass straight through the column. The mRNA is then eluted in a low salt buffer.

5.1.2 Synthesis of double-stranded cDNA

Affinity purification of mRNA is not an essential step in cDNA cloning, because the poly(A) tract can be used to ensure that the mRNA is selectively copied into cDNA (*Figure 5.1*). An excess of oligo(dT) is added as primer and

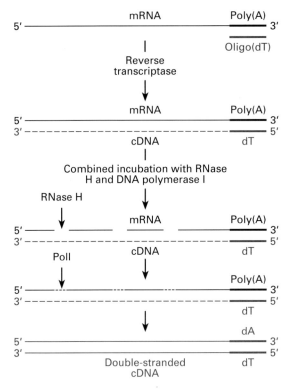

Figure 5.1: *The preparation of double-stranded cDNA. Redrawn from Williams and Patient (1989), with permission from Oxford University Press.*

this anneals to the poly(A) tail. The enzyme reverse transcriptase is used to copy this primer–template complex, to yield a cDNA copy representative of the entire mRNA population. There are a number of available procedures whereby cDNA may be used to generate double-stranded DNA suitable for cloning. A very commonly used method is shown in *Figure 5.1*. The mRNA and cDNA remain annealed after the reverse transcriptase reaction, and the enzyme RNAaseH is added to introduce breaks into the RNA. The resulting nicked molecules act as primer–template complexes for the enzyme DNA polymerase-1, and this produces a double-stranded cDNA copy of the mRNA.

In order to provide cohesive termini, and hence increase the efficiency of ligation into the vector, chemically synthesized restriction cleavage sites (linkers) are often ligated onto the cDNA (*Figure 5.2*). The restriction enzyme site

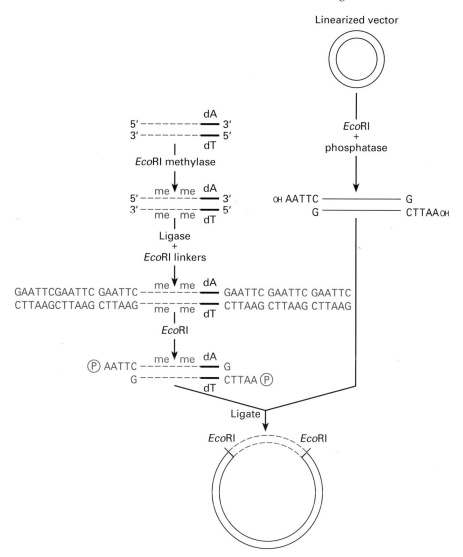

Figure 5.2: *The insertion of double-stranded cDNA into a cloning vector. Redrawn from Williams and Patient (1989) with permission from Oxford University Press.*

present within the linker will occur fortuitously within some cDNAs. Since it is obviously desirable to clone an intact copy of the mRNA, the double-stranded cDNA is treated with a restriction enzyme methylase (Chapter 2) before addition of the linkers. In the example shown in *Figure 5.2*, where *Eco*RI linkers are used for cloning, the target is modified with *Eco*RI methylase. This introduces methyl groups into every *Eco*RI recognition site within the cDNA. After ligation to the methylated target, the linkers are cleaved with *Eco*R1, while the methylated cDNA remains uncleaved. This is just one of the possible routes to cDNA insertion, and there are commercially available kits for this and other methods.

5.1.3 Insertion of the double-stranded cDNA into the vector

The double-stranded cDNA is ligated into the vector of choice, most commonly into an insertion vector such as λgt10. This readily accepts cDNA-sized fragments, has a strong biological selection for the presence of an insert and, because it can be introduced into bacteria by *in vitro* packaging, has a very high efficiency of cloning. This is important if only limiting amounts of mRNA are available for preparing the library. In the case of cDNA libraries there is generally less of a problem with selective loss or under-representation if the library is pooled. Hence it is reasonably safe to use amplified libraries, provided the initial complexity (that is, the number of total clones which were pooled) is high and not too many cycles of amplification are used.

5.2 Screening cDNA libraries

The preparation of the library is most often the least problematical of the steps in isolating a particular cDNA clone (and the problem of course vanishes if a reliable amplified library is available). In contrast, screening a library to identify the required sequence is often very difficult and time-consuming. The scale of the task, and the nature of the strategy used, depends on two factors: the abundance of the required sequence in the cell type used to isolate the mRNA for cloning, and the level to which the cognate protein has been characterized. The following strategies are commonly employed.

5.2.1 Differential screening of cDNA libraries

The cloning strategy which necessitates the least prior information about the sequence to be cloned is differential screening [1]. Its only requirement is that two cell types are available which differ in their level of expression of the desired sequence. The two cell types might be, for example, a hormone-reponsive cell line which is inducible for a particular mRNA and the same cell line cultured in the absence of hormone. First, a cDNA library is prepared from hormone-induced cells and duplicate lifts of the library are made (*Figure 5.3*). One copy filter is screened with radioactively labeled cDNA prepared using mRNA isolated from the induced cells, and the other is screened with labeled cDNA prepared from the uninduced cells. A cDNA clone derived from an mRNA which accumulates only in the presence of the hormone will show a hybridization signal with the probe prepared from the induced cell mRNA, but no signal with the probe prepared from uninduced mRNA.

Differential screening can also be used to identify sequences which show quantitative, rather than qualititative, variation between two cell types. The most abundant mRNA sequences direct synthesis of the most cDNA. Therefore the amount of cDNA which hybridizes to a colony is determined by the abundance of that particular sequence in the starting mRNA population. An autoradiogram obtained in differential screening shows signals which vary in

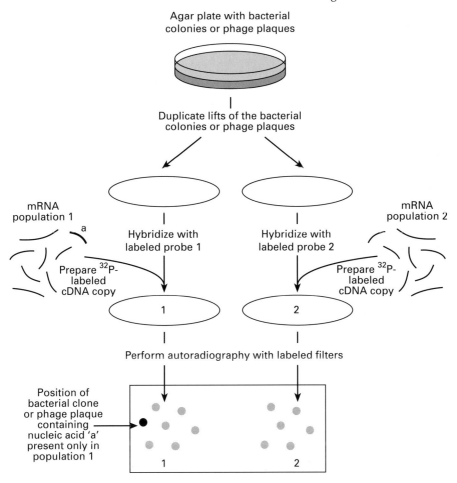

Figure 5.3: *Differential screening of cDNA libraries. Two copies of the recombinant population are made by lifting on to a hybridization membrane. The filters are separately screened with a labeled probe made from each of the two test RNA populations. The resulting autoradiograms are overlaid to search for signals which are present on only one of the two filters. The clone or plaque responsible for this signal is then isolated by eluting the phage or colonies, replating at lower density (for clarity this diagram shows only a few recombinants; in practice a plate might bear 10 000 recombinants) and rescreening.*

strength over several orders of magnitude, reflecting differences in the abundance of individual sequences in the starting mRNA population. By comparing autoradiograms, it is possible to detect differences of as little as three to fivefold in signal strength between duplicate lifts. This is therefore a very useful method of analysis. It allows the patterns of gene expression in two cell types to be compared, and allows genes which show even a small differential expression to be isolated.

A variant of this technique is to remove sequences which are common to the two mRNA populations. A number of strategies have been used for this. In the case described above, the simplest method would be to hybridize an excess of the uninduced cell mRNA to induced cell cDNA (*Figure 5.4*). Sequences which hybridize can then be removed by, for example, column chromatography on hydroxyapatite (a matrix which allows the separation of single-stranded nucleic acids away from double-stranded molecules). The unhybridized, single-stranded cDNA is then used as a substrate for cDNA cloning. The resulting library, usually called a subtraction library, can be analyzed by differential screening using subtracted cDNA and control, unsubtracted cDNA. The great advantage of this method over differential screening is that far fewer clones need to be screened. It is a particularly useful technique for obtaining an enriched source of mRNA for sequences expressed at only a few copies in each cell. However, it is applicable only when there is a qualitative

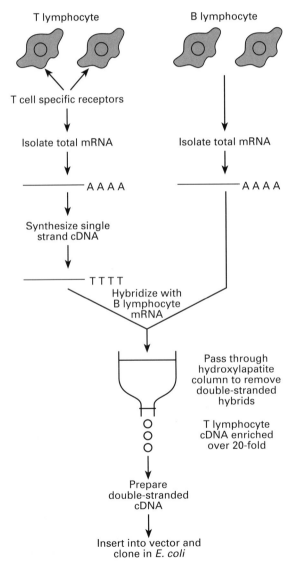

Figure 5.4: *Isolation of the T-cell receptor. This shows the strategy used to obtain an enriched source of cDNA from a T lymphocyte to allow the isolation of the T-cell receptor.*

difference, or a very large quantitative difference, in the abundance of the required sequence in the two mRNA populations (otherwise the required sequence would be lost during the subtractive hybridization).

Subtraction hybridization can also be used to isolate gene products characteristic of a particular differentiated cell type, provided that a closely matched, non-expressing cell type is avaliable. A good example of this approach is provided by the cloning of the T-cell receptor [2]. T-cell receptors are expressed on the surface of mature T lymphocytes and not on B lymphocytes. Both cell types are otherwise very similar and express many of the same gene products. cDNA was prepared from T-lymphocyte mRNA and hybridized with mRNA isolated from B lymphocytes (*Figure 5.4*). The unhybridized, single-stranded cDNA was isolated by hydroxyapatite chromatography and used to construct a cDNA library. Clones for the T-cell receptor were then isolated by screening this library with a cDNA probe prepared from the original T lymphocytes.

Differential screening and subtraction hybridization are very useful techniques for identifying any gene with a differing level of expression in two cell types and for producing a preliminary enrichment for a particular sequence. However, their utility as a means of isolating a specific gene is critically dependent upon the two populations being very similar (aside of course from the sequence or sequences of interest). Two closely related populations are not always available and so other, more generally applicable strategies have been devised. There are two common approaches, oligonucleotide hybridization and expression screening.

5.2.2 *Screening cDNA libraries by oligonucleotide hybridization*

Very powerful techniques are available for determining the amino acid sequences of proteins. If some or all of the sequence of the protein of interest can be determined, an oligonucleotide can then be synthesized chemically which contains the predicted cognate mRNA sequence. This oligonucleotide can then be used as a hybridization probe to screen a cDNA library (see, for example, reference 3). This sounds wonderfully straightforward but, in practice, there is a large element of luck inherent in the method. The amount of work involved depends greatly upon the nature of the known protein sequence. Most amino acids can be encoded by more than one codon (Chapter 1), so that the code is ambiguous when read in the protein to mRNA direction. Leucine, for example, can be encoded by no fewer than six different codons (*Table 1.3*).

If one is fortunate, the protein sequence established will include a short region containing several amino acids encoded by minimally redundant codons, particularly methionine (AUG) or tryptophan (UGG). Even when such codons are present, it is normally necessary to synthesize mixed oligonucleotides containing more than one nucleotide at positions within the chain where codon usage is ambiguous (*Figure 5.5*). The oligonucleotide is hybridized to lifts of the cDNA library under conditions which minimize the amount of non-specific annealing to other genes. However, some non-specific hybridization will normally occur, and many cDNA clones may have to be subjected to

Met Gln Tyr Glu Trp Cys Gln

5' ATG CA$_A^G$ TA$_C^T$ GA$_A^G$ TGG TG$_C^T$ CA$_A^G$ 3'

Figure 5.5: *A multiply redundant synthetic oligonucleotide for use in oligonucleotide screening. Redrawn from Williams and Patient (1989) with permission from Oxford University Press.*

more detailed examination (such as sequence analysis of mini-preparations of DNA) before the required clone is identified.

5.2.3 Expression screening of cDNA libraries

This is a very powerful approach to cDNA cloning when a high-titer antibody is available which is specific for the required protein. A library is constructed in a vector which directs a high level of expression of the proteins encoded by cDNA inserts cloned into it. Most often this is achieved by cloning the cDNA as part of a *lacZ* fusion protein in an expression vector such as λgt11 [4]. The cDNA is cloned into the *lacZ* gene in λgt11 to yield a hybrid protein, containing the region of the *lacZ* gene which encodes the N-terminus of β-gal fused to the eukaryotic cDNA. The *lacZ* gene provides the signals required for efficient transcription and translation in *E. coli*.

After infection by the recombinant phage, cells are exposed to a galactoside which induces expression from the *lacZ* promoter (Chapter 1). The fusion proteins, released from the cell by phage lysis, are lifted on to a filter and detected by binding of the specific antibody (*Figure 5.6*). Plaques which show a positive signal are then purified and recombinant DNA extracted. The final definitive proof that the positive clone is the right one is to show that the cDNA insert has the expected nucleotide sequence. Hence, it is normally necessary to know a part of the amino acid sequence of the required protein.

The series of steps leading from purification of the mRNA, first to the isolation of a cDNA clone and then to a genomic clone, is summarized in *Figure 5.7*. This is given as a guide, but almost every cloning project will incorporate variations, and whole avenues of approach have not been considered. PCR technology can be applied to generate libraries from tiny amounts of starting mRNA and there are very elegant screening strategies based upon cloning cDNA in vectors which direct expression in mammalian cells [5]. The preparation and screening of cDNA libraries is a constantly evolving and vitally important part of genetic engineering technology, and there will doubtless be many further advances. However, present standard procedures make it possible virtually to guarantee isolation of any sequence, provided only that some enzymatic or biological activity can be used as the basis of purification and/or antibody production.

5.3 Identifying genes by positional cloning: cystic fibrosis

All of the gene cloning methods described thus far rely on some knowledge of the encoded protein. Minimally, this would be a knowledge of its distribution in various tissues, so that differential screening could be used. In the best possible case some, or all, of the sequence of the protein would be known.

Overlay phage plaques with
filter impregnated with IPTG

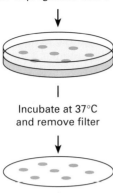

Incubate at 37°C
and remove filter

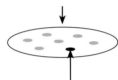

Incubate with antibody and then
with antibody detection system

Position of positive plaque – align filter
with agar and pick plaque

Figure 5.6: *Screening lambda expression libraries. λgt11 contains the lacZ promoter so gene expression is induced with the galactoside IPTG. The induced protein, released from the lysed cells, then sticks to the filter.*

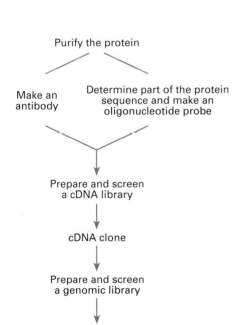

Purify the protein

Make an
antibody

Determine part of the protein
sequence and make an
oligonucleotide probe

Prepare and screen
a cDNA library

cDNA clone

Prepare and screen
a genomic library

Genomic clone

Figure 5.7: *Two common gene cloning strategies. These are two very commonly used gene isolation routes but there are several alternative techniques which can be used if the protein is not easily purifiable.*

There are, however, many inherited diseases where, until the gene is cloned, there is little or no information concerning the defective protein. Cystic fibrosis (CF) is a good example. The gene (known as *CFTR*) was cloned using only a knowledge of its general chromosomal location [6–8]. This provides an excellent example of the enormous power of this procedure of mapping followed by positional cloning.

CF affects approximately 1 in 2000 live births among Caucasians, and has a calculated carrier frequency of 5%. The disease is associated with an abnormal function of chloride channels in apical membranes of epithelia. Electrophysiological studies suggested the existence of an altered ion channel in CF patients. Membrane proteins are notoriously difficult to purify and, before the gene was cloned, there was no direct biochemical evidence to support this hypothesis, nor even to show that altered membrane function was the primary cause of the disease.

The locus was assigned to the long arm of chromosome 7 by identifying genetic markers which tended to co-segregate with the gene in families with several affected children (*Figure 5.8*). The fundamental principle of this pro-

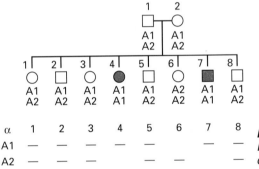

Figure 5.8: *The principle of RFLP mapping of genetic diseases.*

cedure has been described in Section 2.4.3. Two markers that are closely linked in the genome will rarely be separated by recombination. By screening DNA from many families with two or more affected children with a number of RFLP probes that are known to derive from the long arm of chromosome 7, it was possible to identify four RFLPs that showed particularly close linkage to the CF mutation.

Even closely linked loci in man can still be separated by a large amount of DNA. On average, human loci which show only 1% recombination are 1 Mb apart. The linkage defined a region of about 500 kb which was expected to contain the CF locus, and this was cloned by a large-scale program of chromosome walking and jumping (Sections 4.4 and 4.5.4). It was then necessary to identify potential gene sequences within this region which might encode the CF protein. There are a number of general approaches that can be used to solve such a problem.

(a) Direct cDNA cloning. If there is a candidate tissue where the gene is expected to be expressed, a cDNA library is prepared from this tissue, and clones are isolated that hybridize with the chromosomal region known to contain the mutation.

(b) Identification of altered mRNA transcripts in affected individuals. This is sometimes a useful approach but, again, is only possible if there is a candidate tissue in which the gene is expected to be expressed. Detection of a difference by Northern transfer analysis relies on there being a detectable lesion, such as absence of the transcript or a size shift, in one of the avaliable mutant alleles. There are, however, more complex hybridization techniques that are capable of detecting point mutations and very small size shifts.

(c) Inter-specific cross-hybridization. The coding regions of genes are conserved between closely related species, while intergenic regions and introns tend to diverge rapidly. By performing Southern transfer with DNA isolated from various mammalian species (a zoo blot), using different parts of the candidate region as a probe, it is often possible to identify potential coding sequences.

(d) Identification of CpG islands. Many genes have large numbers of CpG dinucleotides in the region encompassing their start sites, often because this area is rich in Sp1 sites (Section 2.4.1). Such sequences are known as CpG islands, because this dinucleotide is under-represented elsewhere in human DNA. Restriction enzymes that contain CpG in their recognition sequences preferentially cut DNA within these islands, so by screening clones with such enzymes, it is sometimes possible to identify genes.

(e) Direct identification of a potential ORF by conventional DNA sequencing. Once candidate coding regions have been identified using the techniques described above, then it becomes feasible to determine their DNA sequence. This might appear straightforward but is, in practice, very difficult. The sequence elements that signal initiation of transcription, gene splicing and polyadenylation are not absolutely conserved. Hence it is not possible to be certain that a predicted ORF really does form part of a gene. The only reliable methods are to perform exhaustive RNA mapping, or to isolate the cognate cDNA clone from an expressing tissue.

In the case of CF, inter-specific cross-hybridization yielded four potential candidate genes. Additional genetic evidence eliminated one of these, and another did not contain an ORF of appreciable size. The third fragment cross-hybridized weakly to other mammalian DNAs, showed a high frequency of CpG islands and the presence of several ORFs, but did not hybridize to a transcript in any of the human tissues tested. The fourth fragment contained CpG-rich regions and hybridized to a cDNA clone in libraries prepared from several tissues. When the cDNA clone was used as a probe in Northern transfers it detected a 6.5 kb transcript in many of the tissues tested. Strong evidence that this was the CF gene came from sequence analysis in affected individuals. Approximately 70–80% of CF-bearing chromosomes in northern Europeans carries a 3 bp deletion that results in elimination of a phenylalanine at position 508 of the deduced protein product. This deletion is never seen on normal chromosomes, and it is associated with the most severe form of the disease. Since the discovery of this first mutation, over 100 other mutations have been described. Some have been seen only in particular ethnic groups, and some seem to lead to a less severe form of CF.

Final confirmation that this was indeed the CF gene came from functional studies. Analysis of the predicted protein sequence suggested it was a transmembrane protein with an ATP-binding site in the cytoplasmic domain (*Figure 5.9*). This structure is consistent with it being an ion-channel protein. Direct

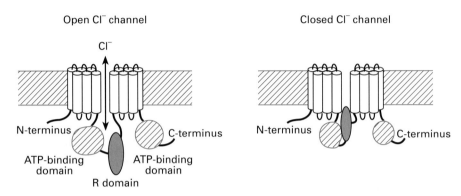

Figure 5.9: *Model of* CFTR *function and regulation. Cystic fibrosis is caused by a defect in chloride ion secretion by epithelial cells. The* CFTR *gene is an integral membrane protein whose function is as an ion transporter. It contains two membrane-spanning domains and a regulatory domain (R) which controls the flow of chloride ions across the membrane in response to cAMP-activated protein kinases. Reproduced from T. Strachan (1992)* The Human Genome, *p. 132, with permission.*

evidence for this derived from experiments where the gene was expressed in cultured cells from CF patients. A wild-type, but not a mutant, form of the cDNA complemented the defect in the chloride channel shown to be characteristic of CF. It seems quite certain therefore that this massive effort, involving several different laboratories and a combination of many skills, has revealed the cause of this long mysterious disease. It has also made possible the construction of specific probes for CF that allow accurate prenatal diagnosis of mutant alleles in families with a history of the disease.

It has also led to proposals to carry out carrier testing in the general population. This has raised a number of ethical and legal issues: who is to be tested, and how are the general public to be made to understand the significance of an individual being a carrier of a defective gene in a recessive disorder? In the United States there has been concern about possible discrimination against gene carriers (who are perfectly healthy) in obtaining health insurance. If these problems can be solved then, hopefully, the development of voluntary carrier screening programs will lead to a significant reduction in incidence and possibly ensure that no couple has a child with CF except as a result of an informed decision to do so.

References

1. Williams, J.G. and Lloyd, M. (1979) *J. Mol. Biol.*, **129**, 19.
2. Hendrick, S.M, Cohen, D., Nielsen, E.A., and Davis, M.M. (1984) *Nature*, **308**, 149.
3. Ullrich A., Berman, C.H., Dull, T.J., Gray A. and Lee, J.M. (1984) *EMBO J.* **3**, 361.
4. Huynh, T.V., Young, R.A. and Davis R.W. (1985) In *DNA Cloning, A Practical Approach,* (D.M. Glover, ed.). IRL Press, Oxford.
5. Seed, B. (1987) *Nature*, **329**, 840.
6. Rommens, J.M., Ianuzzi, M.C., Kerem, B.-S., *et al.* (1989) *Science*, **245**, 1059.
7. Riordan, J.R., Rommens, J.M., Kerem, B.-S., *et al.* (1989) *Science*, **245**, 1066.
8. Kerem, B.-S., Rommens, J.M., Buchanan, J.A., *et al.* (1989) *Science*, **245**, 1073.

Further reading

Collins, F.S. (1991) *Nature Genetics*, **1**, 3.

Collins, F.S. (1992) *Science*, **256**, 774.

Sambrook J., Fritsch, E.F. and Maniatis, T. (1989) *Molecular Cloning, A Laboratory Manual.* Cold Spring Harbor Laboratory Press, New York.

Wicking, C. and Williamson, R. (1991) *Trends Genet.*, **7**, 288.

Williams, J.G. and Patient, R.K. (1989) *Genetic Engineering.* IRL Press, Oxford.

6
CLONING IN HIGHER ORGANISMS

E. coli is the workhorse of genetic engineering but, because chromosome structure and gene expression in prokaryotes and eukaryotes differ so radically, there are some inevitable limitations to its usefulness. Hence methods of introducing and stably maintaining foreign DNA have now been developed for many of the commonly studied eukaryotic organisms. As with bacterial cloning, the process is called transformation, but here the term should not be confused with the phenotypic transformation induced by viral and cellular oncogenes. The ability to transform eukaryotic cells is revolutionizing many aspects of biology.

While the general principles of gene cloning are similar, the eukaryotic cell presents problems and opportunities not applicable to prokaryotes. The most obvious difference is the presence of a nuclear membrane. Fortunately this has not proven a major bar to transformation. When DNA is introduced into eukaryotic cells it readily crosses the nuclear membrane in some as yet undefined manner. Alternatively, DNA can be introduced directly into the nucleus by micro-injection.

The ability to transform eukaryotes has opened up many new possibilities. Thus it is now possible to clone much larger segments of DNA than is possible in *E. coli*, to study their expression in their natural milieu and to modify the phenotype of the host organism.

6.1 Cloning in yeast

The budding yeast, *Saccharomyces cerevisiae*, has many attractions as a host for genetic engineering. The classical genetic approaches, which can best be applied in a haploid cell, have made it the organism of choice for investigating many fundamental processes in the eukaryotic cell. As with *E. coli*, the spin-off from this basic research has been the establishment of a battery of highly sophisticated molecular genetic techniques.

Yeast cells grow with a doubling time of just 90 minutes and can be maintained in either a haploid or diploid state. They have a rigid outer cell wall, which is removed by enzymatic treatment, and the DNA is introduced by incubation in a mixture of calcium chloride and polyethylene glycol. Yeast can be grown in the laboratory in a simple defined medium and strains have been isolated which are incapable of growth in media lacking specific amino acids. These auxotrophic strains contain mutations in the genes responsible for specific amino acid biosynthesis, for example, a TRP⁻ strain is incapable of growing in the absence of added tryptophan. Several of these genes, for example, *TRP1* which encodes an enzyme in the tryptophan biosynthetic pathway, have been cloned. When the *TRP1* gene is introduced into a TRP⁻ cell by DNA transformation it corrects the defect, yielding cells which are rendered capable of growth in the absence of added tryptophan.

The simplest kind of yeast vector [1] contains a selectable marker, such as *TRP1*, joined to a bacterial plasmid vector (*Figure 6.1*). A hybrid, or shuttle, vector of this kind can be used to transform either yeast or *E. coli*. Almost all eukaryotic vectors are of this form, because it is much easier to manipulate DNA by performing genetic engineering in *E. coli*. Transformation efficiencies are high and bacteria grow much more quickly than eukaryotic cells. In a typical experiment a gene under investigation would be ligated into a *TRP1* vector and cloned in *E. coli*. It might then be subjected to further genetic manipulation, such as the introduction of specific mutations, again using *E. coli* as host. Finally, it would be transformed into yeast cells to determine the effects of the mutations on the function of the gene.

When transformed into yeast, such a vector integrates into the yeast genome by recombination. Recombination is mediated by the battery of enzymes which naturally act to exchange DNA between homologous chromosomes. A single homologous recombinational event generates two copies of the *TRP1* gene linked in tandem (*Figure 6.1*). Transformants generated by such an event are very unstable in the absence of selection (that is, when grown in the presence of tryptophan), because recombination between the two tandemly repeated sequences occurs at a very high rate. This results in excision from the genome, and consequent loss, of one of the two copies (*Figure 6.1*). Depending upon the precise position at which recombination occurs, either the mutant or the wild-type gene is retained in the genome. In those cases where the mutant gene is retained, a strain is created with the organization of the wild-type chromosome but carrying the mutation brought in with the vector. This phenomenon of allele exchange thus allows the introduction of specific mutations into precisely targeted regions of the yeast genome.

Yeast is becoming an important organism for the commercial production of medically important proteins, such as viral vaccines. In general, the yield of protein is proportional to the number of copies of the gene present within the cell. Integrating vectors are not generally used for the production of heterologous proteins because they are present in only a single copy in the genome, and are rapidly lost by allelic exchange. Extrachromosomal vectors, based upon a naturally occurring plasmid of yeast called the 2μ circle, are commonly used for this purpose [2]. Yeast cells transformed with recombinants generated in 2μ circle-based plasmids contain 50–100 copies of the vector which is stably maintained within the cell, even in the absence of continued selection.

The class of extrachromosomal vectors of most direct relevance to research in higher eukaryotes are those which allow the cloning of extremely large segments of DNA. Cosmid vectors accept inserts of 45 kb in length and this is generally satisfactory for isolating and analyzing individual higher eukaryotic genes. However, to study the higher order structure of the genome, or to map it by establishing ordered arrays of cloned fragments, demands that very much

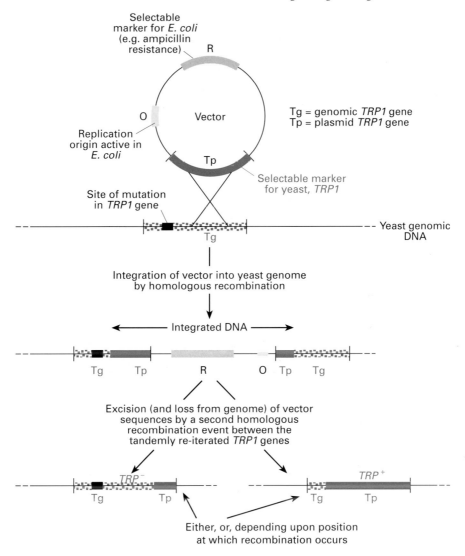

Figure 6.1: *Allele exchange using a yeast integrating vector. The vector carrying an intact* TRP1 *gene integrates by homologous recombination, producing a tandem duplication of the gene. This is an unstable situation (see* Figure 4.13a*) and so one copy of the gene is excised, again by homologous recombination. Depending upon the position at which crossing-over occurs, a cell will inherit the wild-type gene or the mutant gene. Thus, in a fraction of progeny cells, the original defect will be repaired.*

larger fragments be cloned. This has been achieved by the construction of yeast vectors which permit the formation of yeast artificial chromosomes (YACs) [3].

Eukaryotic chromosomes contain a centromere, which is the site of attachment to the spindle, the structure that pulls the chromosomes apart during mitosis and meiosis. Chromosomes are bounded at each end by a telomere, a region that is essential for complete chromosomal replication. Yeast telomeres and centromeres have been isolated by gene cloning. DNA replication in eukaryotic chromosomes is initiated at multiple sites, scattered along the chromosome. Origins of DNA replication in yeast have been cloned by taking advantage of their ability to confer extrachromosomal replication on circular DNA molecules. They are termed autonomously replicating sequences (ARSs). A typical YAC vector contains a centromere, an ARS, two telomeres and two yeast selectable markers (*Figure 6.2*). These elements are present in a

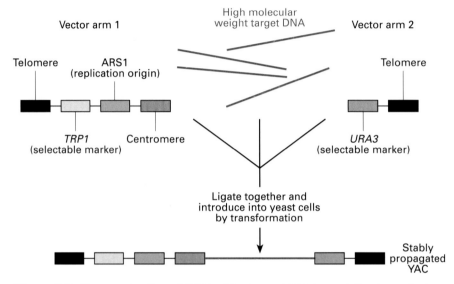

Figure 6.2: *The construction of YACs. The target DNA is partially cleaved with a restriction enzyme and ligated to the two vector arms to produce a YAC clone library. It is essential that the starting DNA be of a very large size and the target is sometimes purified by PFGE before insertion, to remove smaller fragments that would otherwise predominate in the library.*

configuration such that when large fragments of DNA (of up to 500 kb in size) are ligated into them in the manner shown in *Figure 6.2*, a pseudo-chromosome is generated which is stably maintained in yeast as a linear molecule. Such YAC clones can be propagated in yeast in much the same way as a plasmid in *E. coli*. However, for detailed analysis of individual genes it is normally necessary to sub-clone into *E. coli* a fragment from the YAC clone containing the gene of interest.

6.2 Transformation of mammalian cells in tissue culture

Yeast is of great value as an organism for basic research, for producing heterologous proteins and as a vehicle for cloning large segments of DNA. For some purposes, however, it is essential to be able to transform mammalian cells. While yeast and mammals share many basic properties, there are important differences in the precise mechanics of gene expression. Mammalian transcription signals are not generally recognized in yeast, mammalian RNA transcripts are not spliced and mammalian proteins are not correctly glycosylated. Therefore basic research into the control of higher eukaryotic gene expression and the commercial production of some proteins, where secondary modification is considered important, have required the development of transformation methods for mammalian cells. These techniques enable us to investigate the properties and differentation of mammalian cells representative of most cell lineages in much the same way that single-celled organisms, such as yeast, are studied.

The essential background to achieving transformation was the establishment, over many years of painstaking work, of methods for growing mammalian cells in the laboratory. Because they are normally protected from the rigors of the world, mammalian cells are much less robust than yeast cells. They require a carefully balanced growth medium and strict control of pH and temperature. Even when provided with these conditions, cells freshly isolated from a tissue have only a finite lifetime in culture; they will grow and divide for only a limited number of generations. However, in cultures derived from some tissues, mutant cells arise which have acquired effective immortality in culture. These cell lines are used as the host cells for transformation.

The changes which accompany immortalization mimic events occurring when a cell is transformed by an oncogene and, in those cases where cell lines do not arise spontaneously upon prolonged passage, tumors are an important source of novel cell lines. Although tumor cells often lose the ability to enter their normal pathway of differentiation, it is sometimes possible to identify cell lines which will replicate under one set of culture conditions and differentiate under another. For example, cell lines derived from mouse teratocarcinomas, such as the F9 embryonal carcinoma cell, will differentiate to form visceral endoderm (one of the tissues produced early during embryogenesis) when deprived of a surface for attachment and exposed to retinoic acid [4].

Where a naturally occurring tumor cell line is not available, it is sometimes possible to generate cell lines by exposure of appropriate primary cultures to oncogenic viruses. Thus, in many genetic studies there is a particular need for a renewable supply of DNA from key individuals. This is especially important in studies of those genetic diseases in which the affected people do not survive for a long period after diagnosis. To overcome this problem, and to avoid resampling, it is essential to establish an immortalized cell line. The Epstein–Barr virus (EBV) is able to integrate into human B lymphocytes and transform them into a permanent cell line. To establish a lymphoblastoid cell line, a mixed population of T and B lymphocytes is exposed to EBV and cultured for 3 to 4 weeks. EBV transformation does not alter the karyotype, and the cells grow very rapidly and can be stored frozen for extended periods of time.

6.2.1 Transformation of mammalian cells with integrating vectors

Mammalian tissue culture cells readily take up DNA when subjected to electric shock (electroporation), or when the DNA is co-precipitated with calcium phosphate and overlaid on to a monolayer of cells growing on a plastic Petri dish. Depending upon the particular method used to introduce the DNA and the nature of the cell line, a large fraction of cells may take up the vector and transiently express genes contained in it. This phenomenon of transient expression is a useful method of studying gene regulation, because cells can be analyzed very soon after exposure to the DNA.

Only a very few cells go on to produce stable transformants, where the introduced DNA has integrated into the chromosome. It is therefore necessary to have some means of selecting for transformant clones. The most commonly used method of selecting a stable transformant is to include within the DNA molecule a gene which confers resistance to a toxic drug, such as the protein synthesis inhibitor G418. This drug is similar in structure to the antibiotic neomycin. A naturally occurring bacterial resistance gene (sometimes termed the *neo* gene) acts by phosphorylating, and hence inactivating, neomycin and G418. In order to direct expression of the *neo* gene in a mammalian cell, its coding region has been fused to transcription and RNA processing signals which function in eukaryotic cells. A typical G418 resistance vector, pSV2–*neo* [5], contains a promoter and RNA processing signals from the SV40 tumor virus flanking the drug resistance gene (*Figure 6.3*).

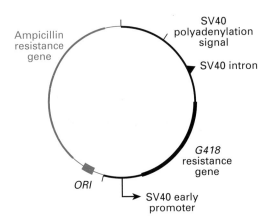

Figure 6.3: *A mammalian integrating vector. This prototypic vector contains transcription and RNA processing signals from the animal virus SV40 driving expression of a bacterial neomycin resistance gene. This confers resistance to G418, the related drug which is active in animal cells. It is a shuttle vector for* E. coli *so it also contains an ampicillin resistance gene and an origin of replication (ORI).*

If two different DNA molecules are introduced together into cells, then a high proportion of stably transformed clones will contain both sequences. This technique of co-transformation is very useful because, by mixing a DNA sequence of interest with a G418 resistance vector prior to precipitation with calcium phosphate, the DNA can be introduced into a cell without the necessity of first cloning it into a vector.

When mammalian cells are stably transformed with a vector such as pSV2-neo, the DNA integrates at one or more regions within the genome by a process of non-homologous recombination. Homologous recombination does also occur (see Section 6.4), but very much less frequently than in yeast, where almost every recombinational event is homologous. The incoming vector DNA may be interrupted at any point along its length, and the sites of integration within the genome also appear to be more or less random. This can pose problems. The metazoan genome is subject to a hierarchy of control not apparent in unicellular organisms. Whole regions of the genome, containing genes which are not expressed in a particular cell type, may be held in an inactive state that is heritably maintained within that cell lineage (see Section 1.8). This repression mechanism can operate to inactivate foreign genes integrated within such regions. Such position effects can be circumvented by using extrachromosomal vectors based upon viral replicons such as the EBV.

6.2.2 Transformation of mammalian cells with retroviral vectors

Retroviral particles contain a genome comprising single-stranded RNA. In infected cells the RNA is copied into a circular, double-stranded DNA which integrates into the cellular genome to form a provirus. The provirus contains long terminal repeats (LTRs), virally derived sequences a few hundred nucleotides long (*Figure 6.4*). Retroviral vectors contain the two LTRs flanking a selectable marker, such as a G418 resistance gene, and other viral sequences

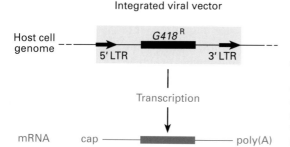

Figure 6.4: *A retroviral vector. In such a vector expression of G418 resistance, and integration into the genome, are both directed by retroviral DNA sequences.*

necessary for replication and packaging of the RNA into infectious particles. They also contain bacterial plasmid sequences, and DNA to be introduced into a mammalian cell is first cloned into a site between the LTRs using *E. coli* as a host. The recombinant vector is then introduced, by transformation and selection for G418 resistance, into mammalian cell lines that contain the viral genes encoding the proteins necessary to package phage RNA into infectious particles. Because the DNA insert in the vector is flanked by the viral LTRs and packaging sequences, it is transcribed, packaged and exported from the cell. These pseudo-viral particles are then used in a second transformation step and, because viral infection is a highly efficient process, most or all of the cells become transformed. This single step, high efficiency transformation of cells that may be refractory to other methods of introducing DNA, constitutes one of the great advantages of retroviral vectors.

A further advantage is that integration of retroviral vectors is a relatively controlled process. In contrast to transformation with randomly integrating vectors, which normally yields cell clones containing multiple gene copies, retroviral vectors integrate into the host cell genome as a single copy. Also, the virus always integrates with the LTR sequences at its boundaries (*Figure 6.4*). They are not good vectors for studying tissue-specific expression, because the enhancers can lead to inappropriate expression of sequences cloned between them, but they are invaluable if the aim is to obtain constitutive expression. For this reason they are considered to have great potential as vectors to correct human genetic defects in constitutively expressed genes.

6.3 The creation of transgenic animals

Cultured mammalian cells are ideal for many purposes in cell and developmental biology, and great advances have been made using them. Sometimes, however, it is essential to study the whole organism. This is obviously necessary for those genetic diseases where a lesion affects many tissues and organs. It also holds true when the behavior of cells in culture does not faithfully reflect the behavior of the analogous cells *in vivo*. Mammalian development is a complex process, involving multiple interactions of cell with cell and of cell with substratum. It is perhaps not surprising, therefore, that development *in vitro* sometimes fails completely to mirror development *in utero*.

The laboratory mouse was the testing ground for the establishment of 'whole mammal' genetic engineering, but similar procedures have since proven successful for other, commercially important, species. Newly fertilized eggs are collected, prior to fusion of the male and female pronuclei. A tiny volume of DNA solution is sucked up into a very fine glass pipet and injected into the male pronucleus, which can be distinguished from the female pronucleus because it is smaller in size. The eggs are then implanted into pseudo-pregnant foster mothers, generated by mating with vasectomized males.

The procedure is technically demanding but, in skilled hands, a proportion of injected eggs survive to term and will contain the injected DNA. Progeny mice are tested for the presence of this DNA by removing a portion of the tail and performing Southern transfer (a procedure normally termed a tail blot) or PCR analysis. Integration of the DNA into the genome of cells within the embryo is a poorly understood process, but it seems to occur early during embryonic development. A proportion of the animals resulting from injection are mosaic; the various tissues within the animal contain differing amounts of the injected DNA, and it is absent from some tissues. It is normally necessary to establish a line of mice containing the gene of interest in all tissues. Fortunately, injected DNA integrates into the germ cells in a high proportion of mice. By mating such mosaic mice to non-injected animals and analyzing tail blots of the progeny, it is possible to identify those containing the integrated gene. The parent in such transgenic lines of animals will pass on the injected gene in a simple Mendelian fashion (*Figure 6.5a*), and is said to display germ-line transmission. The DNA is injected as a linear fragment which forms head to tail concatemers that normally integrate at apparently random locations within the genome. Again, the gene may be susceptible to position effects, so that different mouse lines containing the same gene may show very different levels of gene expression.

Transgenic mice have many important uses. They can be employed as an assay system, to delineate the DNA sequences necessary for tissue-specific gene expression in those cases where there is not an appropriate tissue culture system. They can be used to generate mouse models of genetic diseases. If, for example, an oncogene is placed next to a promoter which directs tissue-specific gene expression, it can lead to the formation of specific tumors. They can also be used to perform random insertional mutagenesis to search for genes important in mouse development. In a surprisingly high proportion of cases integration of the transgene occurs in a region of the genome which inactivates an important gene. Such a mutation (if recessive) will not manifest itself in a transgenic line maintained by mating to a wild-type mouse (*Figure 6.5a*), because the normal allele will always be present and will provide the missing function. It can, however be revealed by crossing two sibling mice, heterozygous for the transgene (*Figure 6.5b*). One-quarter of the progeny of such a cross will be homozygous for the mutation and if, as is often the case, it is a gene required during development, these embryos will die *in utero*. The beauty of this technique of mutagenesis is that the vector DNA used to generate the transgenic mouse will normally be located somewhere in the vicinity of the mutated gene. It is therefore a relatively straightforward matter to prepare a genomic clone bank from mutant mice, to screen for the integrated DNA, and hence to clone the essential gene.

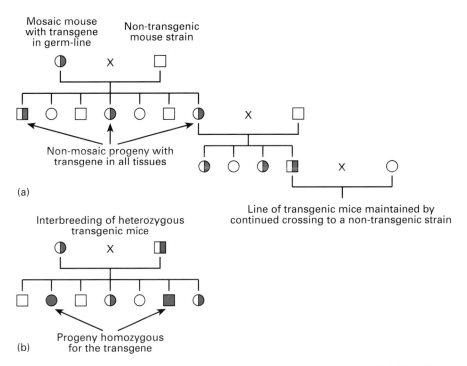

Figure 6.5: *Maintenance and analysis of transgenic mouse strains. (a) This figure shows how a line of transgenic mice can be maintained by crossing to a non-transgenic strain and monitoring progeny for the presence of the transgene to determine which mice are carriers. (b) If the transgene is recessive, its effect can be determined by mating together two transgenic mice. Assuming that the transgene is not essential for fetal development, 25% of the progeny will be homozygous for the transgene.*

6.4 Gene targeting by homologous recombination

Cloning has allowed the isolation of mammalian genes involved in a wide spectrum of biological processes. Even though the biochemical activity or structural role of the gene product will normally be known prior to cloning, it is often very valuable to determine the phenotypic consequences that result if it is absent. In lower eukaryotes, such as yeast, this has been routine for many years, using either allele exchange or simple gene disruption. The latter procedure is now applicable to mammalian cells. This ability to perform 'reverse genetics', to clone a gene and then determine the consequences of its ablation, promises to revolutionize mammalian cell and developmental biology.

The technique is much more time-consuming and demanding than in yeast because the frequency of homologous recombination in mammalian cells is orders of magnitude lower than in yeast, and because mammals are diploid organisms. The latter problem could, in principle, be overcome by performing disruption in transgenic mice, mating together two transgenic animals and identifying the 25% homozygous mutant progeny (*Figure 6.5b*). However, the frequency of homologous recombination in transgenic mice is probably as low as that observed for mammalian tissue culture cells, in the order of one homologous event to 1000 non-homologous events. Thus it would be totally impractical to generate sufficient numbers of transgenic mice to identify homologous recombinants because of the length of time required.

The solution to this problem is to produce mutant mice by creating chimeras with embryonal stem (ES) cells, rather than by direct micro-injection of DNA into eggs [6, 7]. ES cells are embryo-derived cells that are pluripotent, that is, they have the ability to participate in normal development and to differentiate into every mouse cell type. If ES cells are introduced into an embryo, a chimeric animal is produced. This is a mosaic, containing both ES and normal cells. Since the germ cells will also be a mixture of ES and normal cells, chimeric animals produce a fraction of offspring derived from ES gametes. ES cells are susceptible to DNA-mediated transformation in culture, using integrating vectors, and such transformed cell lines retain the ability to chimerize with normal cells. This provides an alternative method for generating lines of transgenic mice (*Figure 6.6*).

The importance of the procedure is that hundreds or thousands of ES cell lines, each the product of a different transformation event, can readily be generated *in vitro*. These can then be screened, by Southern transfer or PCR, to identify the rare cases in which the gene has integrated by homologous recombination (*Figure 6.7*). There are also now genetic methods to increase the frequency of cell clones in which homologous recombination has occurred [8]. Such a cell line will not display a mutant phenotype which is recessive because the cells are diploid and so retain a normal copy of the gene on the non-disrupted chromosome. However, the effect of homozygosity for the mutation can be determined by breeding homozygotes from the original heterozygous transgenic mice.

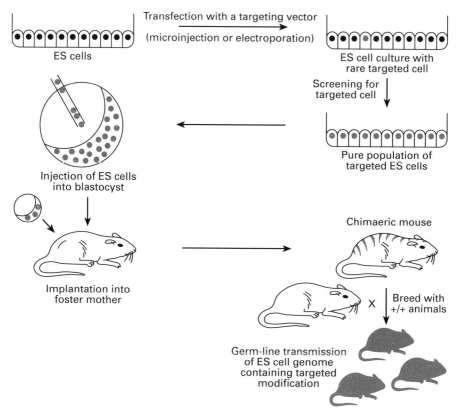

Figure 6.6: *Generation of transgenic mice using ES cells. This is an alternative route to producing a transgenic mouse. Since ES cells are cultured cells which can undergo spontaneous mutation during passage, it is essential that the cells be carefully maintained, otherwise they will not contribute the germ line of the progeny mice.*

6.5 Hox *genes,* Pax *genes and the new genetics*

A good example of the use of targeted gene disruption is provided by a recent study of the phenotypic effect produced by inactivating the mammalian *Hox-1.5* gene [8]. This same study also illustrates how mice can be used to create a model of a human disease.

Homeotic genes were first identified in the fruit fly *Drosophila melanogaster* as genes which, when mutated, cause aberrant development. The best characterized of these genes, those which help to specify the pattern of differentiation along the antero-posterior axis, are transcription factors. They determine the identity of the various segments by activating genes that direct formation of the structures characteristic of each segment. Mutation in various of these genes can yield remarkably aberrant adult flies with, for example, an additional winged segment, giving four wings instead of the normal two, or with a leg-bearing segment on the head. The main *Drosophila* homeotic genes are organized into two clusters and for reasons which are not yet clear, those genes which are active in the most anterior segments of the fly are located nearer the 3′ end of the clusters, and genes that are active in the more posterior

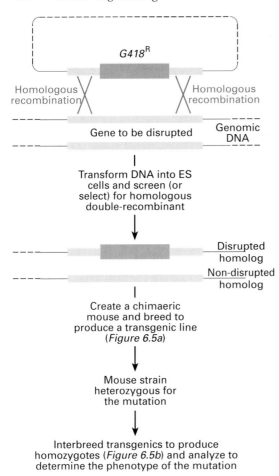

Figure 6.7: Homologous gene disruption in ES cells. A series of cloning steps in E. coli are used to generate a construct in which the 5′ proximal and 3′ proximal parts of a gene flank a G418 resistance gene. If this inserts by homologous recombination, one of the allelic copies of the gene is inactivated. The phenotype of the mutation is then determined by creating a transgenic mouse and breeding to create a homozygote.

segments are located nearer the 5′ end. There is an almost perfect inverse correlation between position of a gene relative to the 5′ end of its respective cluster and its domain of expression within the fly (*Figure 6.8*).

The DNA-binding domain of homeotic genes, the homeobox, is present, in a more or less conserved form, in all metazoan organisms so far studied. Homeobox genes (homeogenes) have been isolated from several different mammals, including man, and they are proving to be the Rosetta stone of mammalian developmental biology. Just as in the fruit fly, they play a central role in regulating formation of the antero-posterior axis during embryonic development. Remarkably, given the enormous evolutionary separation and radically different body plans of mammals and fruit flies, homeogenes that specify particular segments in flies have exact counterparts in mammals.

In the mouse there are four homeogene clusters, found on separate chromosomes, that contain homologous genes in equivalent relative positions. These are believed to have arisen by multiple duplications during evolution from a single cluster, with a structure similar to that obtained by aligning the two *Drosophila* clusters end to end (*Figure 6.8*). Part of the evidence for this hypothesis is that the almost every mammalian homeogene is in an identical relative position to its nearest *Drosophila* homolog.

The sequence homology, and conservation of relative order along the

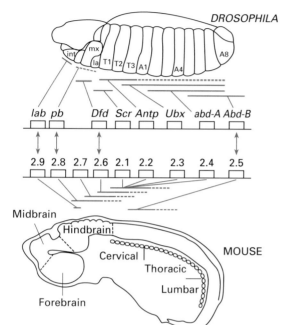

Figure 6.8: Drosophila *and* mouse homeobox genes. Reproduced from [9] with permission from Cell Press.

chromosome, between *Drosophila* and mammalian homeogenes most probably reflects conservation of function. While adult mammals are not externally segmented, during embryonic development the hindbrain is visibly and functionally divided into segments called rhombomeres. The domains of expression of the mammalian homeobox genes in the rhombomeres of the embryonic brain are related to their positions within the gene cluster, in a very similar fashion to their *Drosophila* counterparts. This observation adds confidence to the idea that the homeogene clusters have been conserved during evolution and are playing fundamentally similar roles in flies, mice and men. The downstream target genes of these and other homeobox genes which are believed to direct processes such as limb formation, are sufficiently different to ensure that men have arms and flies have wings.

Drosophila can be bred in enormous numbers so that very sophisticated genetic techniques can be used to analyze its development. These are not easily applicable to the mouse and, whilst conservation of gene order, sequence and pattern of expression are strongly indicative of functional homology, they do not constitute definitive proof that the mammalian homeobox genes play analogous roles to those of *Drosophila*. This proof has come from reverse genetics, which might best be described as 'first catch your gene and then study its function'.

Targeted disruption of the mouse homoebox gene *Hox-1.5* [8] was achieved by transfecting into ES cells a construct with the structure shown in *Figure 6.7*. It contains a G418 resistance gene, flanked on each side by sequences derived from the *Hox-1.5* gene. This generated a number of G418-resistant cell lines, each originating from a different transformation event. Individual clones were then screened to find one in which two homologous recombination events had occurred, as shown in *Figure 6.7*. The effect of this double crossover was to replace the *Hox-1.5* gene on one of the two homologous chromosomes with an interrupted, inactive copy of the gene. These cells were introduced into a mouse embryo to create a chimera which was then re-implanted in a pseudo-pregnant mother.

The ES cells which were used to create the disruptant contained a mutation in the *agouti* coat color gene, so that chimeric mice were readily identified. Breeding from these chimeric animals yielded mice which were entirely agouti in coat color, showing that some of the re-implanted ES cells had contributed to the germ line. By inter-crossing these animals and analyzing the genomic DNA of the progeny by Southern transfers, mice homozygous for the disrupted *Hox-1.5* gene were obtained. These embryos displayed severe defects including athymia and aparathyroidy, and died a few hours after birth. A more detailed investigation revealed that the pattern of malformation closely resembled that observed for human Di-George Syndrome. Di-George syndrome itself is probably not caused by disruption of the human equivalent of *Hox-1.5*, because it maps to the wrong chromosomal location, but clearly the pathological sequence has many features in common with events in the transgenic mice.

The phenotype obtained by disrupting the *Hox-1.5* gene is complex, but it can be partially correlated with the pattern of *Hox-1.5* expression. Redundancy of function between the homologous *Hox* genes may explain why the correlation is not simple. When *Hox-1.5* is disrupted, another homeobox gene may be able to fulfil some of its functions. It probably also reflects the fact that disruption is a very unsubtle form of genetic alteration. A more informative mutation would be one that altered the pattern of expression of a homeobox gene, such that it was expressed in regions of the embryo where it was normally silent. Such ectopic expression can be obtained by linking the coding region of a gene to a promoter from another gene to create a gene fusion.

When the mouse *Hox-1.1* gene is coupled to the constitutively expressed β-actin promoter it produces a remarkable phenotype [10]. The mice have an extra cervical vertebra and variations in the most anterior units, consistent with a posterior to anterior transformation. This change in vertebrate body plan, produced by ectopic expression of the mouse homolog of a *Drosophila* homeotic gene, provides a graphic illustration of the fundamental similarity in the body-building processes between different metazoan organisms.

Another excellent example of the interplay between fly, mouse and human genetics is provided by the *Pax* genes. This class of genes was originally identified in the *Drosophila* mutant *paired*. Using the DNA-binding domain (paired box) of the *paired* gene as an initial probe, a family of *Pax* genes has been isolated from the mouse. One of the genes, *Pax-3*, was found to be expressed in the developing nervous system of the mouse embryo and located in the proximal portion of mouse chromosome 1. A mouse mutation, *Splotch*, maps to this approximate region and has defective function of the embryonic neural crest. It has now been shown that *Pax-3* is the *Splotch* gene [11]. A mutation in another *Pax* gene, *Pax-6*, causes the *small-eye* mutation in mice [12].

Waardenburg's Syndrome (WS1) is an autosomal dominant condition in man characterized by deafness and changes in pigmentation. These symptoms point to defective function of the embryonic neural crest. WS1 has been mapped to a region of chromosome 2 that is homologous to the region of mouse chromosome 1 containing the *Splotch* gene. The human homolog of the *Pax-3* gene has now been cloned, and sequence analysis shows that there are mutations in this gene in patients affected with WS1 [13,14]. Thus *Splotch* and the gene for WS1 are homologous *Pax* genes.

This illustrates the enormous power of having a map. *Pax-3* was tested as a candidate for the *Splotch* and WS1 mutations mainly because they were found to occupy the same part of the chromosome. It will be also clear from these two examples that the discovery of the homeobox and the paired-box genes in *Drosophila* has given a tremendous boost to mammalian developmental biology and to the study of inherited developmental defects in the human. It is a classic example of the importance of studying fundamental biological processes in a system where sophisticated genetics are available.

References

1. Struhl, K., Cameron, J.R. and Davis, R.W. (1979) *Proc. Natl. Acad. Sci. USA*, **16**, 1035.
2. Biggs, J.D. (1978) *Nature*, **175**, 104.
3. Burke, D.T., Carle, G. and Olson, M. (1987) *Science*, **236**, 806.
4. Hogan, B.L.M., Taylor, A. and Adamson, E. (1981) *Nature*, **291**, 235.
5. Southern, P.J. and Berg, P. (1982) *J. Mol. Appl. Genet.*, **1**, 327.
6. Martin, G.R. (1981) *Proc. Natl. Acad. Sci. USA*, **78**, 7364.
7. Evans, M.J. and Kaufman, M.H. (1981) *Nature*, **292**, 154.
8. Chisaka, O. and Capecchi, M. (1991) *Nature*, **350**, 473.
9. McGinnis, W. and Krumlauf, R. (1991) *Cell*, **68**, 283.
10. Kessel, M., Balling, R. and Gruss, P. (1990) *Cell*, **61**, 301.
11. Epstein, D.J., Vekemans, M. and Gros, P. (1991) *Cell*, **67**, 767.
12. Hill, R.E., Favor, J., Hogan, B.L.M., Ton, C.C.T., Saunders, G.F., Hanson, I.M., Prosser, J., Jordan, T., Hartie, N.D. and Heyningen, V. (1991) *Nature*, **354**, 522.
13. Tassabehji, M., Read, A. R., Newton, V. E., Harris, R., Balling, R., Gruss, P. and Strachan, T. (1992) *Nature*, **355**, 635.
14. Baldwin, C. T., Hoth, C. F., Amos, J. A., da-Silva, E. O. and Milunsky, A. (1992) *Nature*, **355**, 637.

Further reading

Capecchi, M.R. (1989) The new mouse genetics: altering the genome by gene targeting. *Trends Genet.*, **5**, 70.

Hill, R. and Van Heyningen, V. (1992) Mouse mutations and human disorders are paired. *Trends Genet.*, **8**, 119.

Kingsman, S.M. and Kingsman, A.J. (1988) *Genetic Engineering*. Blackwell Scientific Publications, Oxford.

Krumlauf, R. (1991) The HOX gene family in transgenic mice. *Current Opinion Biotech.*, **2**, 796.

McGinnis, W. and Krumlauf, R. (1991) Homeobox genes and axial patterning. *Cell*, **68**, 283.

Miller, J.H. and Carlos, M.P. (1987). *Gene Transfer Vectors for Mammalian Cells. Current Communications in Molecular Biology.* Cold Spring Harbor Laboratory Press, New York.

Sclessinger, D. (1990) Yeast artificial chromosomes; tools for mapping and analysis of complex genomes. *Trends Genet.*, **6**, 248.

Sedivy, J.M. and Joyner, A.L. (1992) *Gene Targeting.* W.H. Freeman, London.

7
FUTURE PROSPECTS

7.1 Using molecular biology to understand and cure human diseases

One of the great triumphs of molecular biology over the last 20 years has been the identification of the causative agents for many previously mysterious diseases. This is strikingly illustrated by the enormous progress that has been made in understanding the cellular changes that cause cancer.

7.1.1 Cancer

Many genes have been identified which, when present in a mutant form, cause cells to become tumorigenic. Where their function is known, the normal counterparts of such proto-oncogenes have often been found to play a role in the control of cellular proliferation. They may be changed into a functional oncogene by mutations that alter their biochemical properties or that result in a change in their level of expression. Thus a point mutation that alters the RAS gene into an oncogene changes the rate at which it hydrolyses GTP, the nucleoside triphosphate that regulates its activity. The mutation that alters the MYC transcription factor into an oncogene in some forms of leukemia is a chromosome rearrangement that probably results in its aberrant expression.

In order to render primary tissue culture cells oncogenic it is often necessary to introduce two different oncogenes that act co-operatively to induce cellular transformation. The RAS and MYC oncogenes form such a co-operating pair [1]. The fact that multi-hits appear to be necessary for oncogenic transformation is believed to explain the fact that some forms of naturally occurring cancers, such as bowel cancer, are progressive diseases. It also produces a satisfying explanation for the fact that the frequency of occurrence of cancer shows a strikingly non-linear dependence upon age.

The RAS and MYC oncogenes were isolated because of their ability to transform cells in culture. Evidence that they actually play a role in naturally occurring cancers derives from the presence of a potentially oncogenic form of each of these genes in human tumors. Because the point mutation in the RAS oncogene normally occurs within the same codon [2], it is possible to use oligonucleotide probes, in Southern transfer or in PCR, to detect a change at this position. The re-arrangement around the MYC gene can readily be mapped using Southern transfer.

RAS and *MYC* are 'dominant' oncogenes; when the oncogenically activated form of the gene is introduced into an immortalized cell line it becomes tumorigenic, despite the fact that a normal copy of the gene is also present. There are also recessive oncogenes or tumor suppressor genes, such as the retinoblastoma (*RB*) gene, that appear to act as suppressors of cell proliferation [1]. Only if both copies of the gene are inactivated by mutation do cells become tumorigenic. Biochemical studies have shown that in normal cells the *RB* protein is complexed with several proteins that are potential activators of cellular proliferation [3]. It is assumed that these proteins are rendered inactive when bound to *RB* and that deletion of the *RB* gene, or mutations that inhibit its binding to other proteins, are oncogenic because these associated proteins then become constitutively active. *RB* is not alone in these properties. The *p53* gene encodes another cellular protein that acts as a tumor suppressor, apparently by a similar mechanism. Remarkably, point mutations in the *p53* gene have been found in almost half of the human cancers characterized so far, including cancer of the colon, breast, lung and liver [3].

These results have major implications for the diagnosis of cancer. It should be possible to design diagnostic probes that will allow identification of the lesions present in particular cancers, as they first present and during their progression. These can be used to give a much more coherent and meaningful categorization of tumor type, based upon the oncogenes that are active within the cell rather than upon the cell's histological appearance. A body of information can thus be built up describing the likely prognosis for the various categories, and this could be of great value in planning therapy.

A knowledge of oncogene re-arrangements is already being used in the treatment of hematopoietic cell cancers, such as leukemias and lymphomas. These are treated by whole body irradiation, or with massive doses of selectively cytotoxic drugs. The aim is to destroy all the hematopoietic stem cells in the bone marrow population, so that it can be repopulated with healthy cells. This is normally done by grafting the marrow from a compatible donor and keeping the patient under immunosuppressive therapy, in order to avoid a graft versus host reaction. This type of surgery is extremely difficult and the success rate is low. A graft versus host reaction can occur at any time, sometimes years after the transplantation. It is therefore preferable to use the patient's own marrow. The explanted marrow is purged of cancer cells by chemotherapy or irradiation. Provided a genetic alteration in the cancer cells is known, it is possible to design a PCR probe that will distinguish them from the non-malignant cells. Such a genetic difference will often be provided by the rearrangement of DNA that led to the cells becoming oncogenic. Under appropriate conditions the DNA from a single cancer cell will produce a detectable amount of PCR product. When it is clear that no cancer cells survived the purging, the marrow can be safely re-implanted.

A similar approach can be used to analyze blood for residual cancer cells after the less extreme drug treatments used to treat some forms of leukemias. It is then possible to assess the efficacy of the therapy and predict the likely extent of remission. By genetically typing the cells when the disease is first diagnosed, it is also possible to discriminate between a blastic crisis (a recurrence of the disease due to sudden expansion of the previously existing cancer cell population) and a new (primary) manifestation of the same disease.

We have chosen to discuss cancer in some detail because there have been such great strides in our understanding. However, the power of molecular biology is such that most of the important human diseases are being studied using its methods. While it is beyond the scope of this book to discuss them individually another excellent example of the revolution brought about by molecular techniques is in the understanding and treatment of infectious and parasitic diseases.

7.1.2 Infectious and parasitic diseases

There are many cases where the causative agent for an infectious disease is known but where there is a limitation to studying it, and to creating a vaccine to counteract it, because it is impossible to purify sufficient biological material from the desired stage in the organism's life cycle. Such diseases are now being attacked by a common strategy. The first step is to identify, by molecular cloning, a surface protein which is expressed at a susceptible stage in the organism's life history. Large amounts of this protein are then prepared using an expression vector and tested for its efficacy as a vaccine. Alternatively, a region of the protein which is strongly antigenic is identified and a short peptide that encompasses this region is synthesized chemically. This approach is being applied to a whole variety of agents including bacteria, viruses and parasitic organisms.

A promising further refinement is to create multi-potent immunogens using vaccinia virus vectors [4]. A vaccinia vector will accept large inserts, so that genes encoding many different viral, bacterial or parasitic antigens can be inserted into it and used in immunization. This is a particularly attractive prospect for use in developing countries, where low cost, low technology therapies are essential and where so many debilitating infectious diseases are endemic.

7.1.3 Anti-sense approaches

Immunologically based therapies are presently the most promising short-term applications of molecular biology to medicine but other approaches are also being pursued. Diseases which subvert the biology of individual cells, such as cancer or viral infection, offer a powerful potential target for attack, in that specific gene expression is required to induce the diseased state. If one or more of these mRNAs can be specifically destroyed then the disease can, in principle, be cured. While not optimal for the process, the conditions of ionic strength and temperature within a cell are permissive for nucleic acid hybridization. If, therefore, the complementary strand to a specific mRNA is expressed within a cell it will anneal with mRNA to form a hybrid (*Figure 7.1*). Such a hybrid is often rapidly degraded within the nucleus or the cytoplasm of the cell and is, in any event, untranslatable. Provided, therefore, that it is present in a sufficient excess over its cognate mRNA, such an anti-sense RNA will block the accumulation and/or expression of the target mRNA. High level

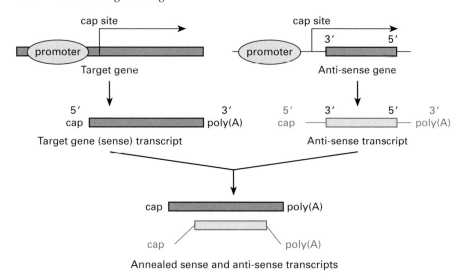

Figure 7.1: *Anti-sense therapies.*

of expression can be achieved by introducing an anti-sense copy of the coding region of a gene into cells in multiple copies under the control of its own promoter or by placing it under the control of a more highly active heterologous promoter.

Anti-sense inhibition has been used to create plant strains which are resistant to viral diseases and it seems possible that a similar approach will be used to create transgenic animals which are resistant to commercially important viral diseases. In the case of human viral diseases, transgenesis is not an applicable approach and considerable effort is being devoted to finding systemically active reagents based on the same principle. Fortunately, only a small piece of complementary nucleic acid is needed to block translation. Hence, one exciting possibility is to use anti-sense oligonucleotides which are chemically modified so as to be stable within the cytoplasm of cells [5]. Nucleic acids are very hydrophilic and so will not pass through the cell's plasma membrane. It will therefore be necessary to find a way to circumvent this major problem, perhaps by coupling them to a ligand which is efficiently internalized in the target cell or by introducing them into liposomes, artificial membranes which can fuse with the plasma membrane and expel their contents into the cell.

For accessible tissues, an interesting alternative is simply to paint the anti-sense oligonucleotide on to the exposed tissue. A recent demonstration [6] showed that this approach was surprisingly effective. If the inner epithelium of a rat's carotid artery is removed by balloon angioplasty, smooth muscle cells proliferate and threaten to occlude the artery. A similar phenomenon in man is a major cause of long-term failure of coronary arterial bypass grafts. The cell proliferation could be very largely suppressed by painting the exposed arterial wall with a quick setting gel containing an anti-sense oligoncleotide to the product of *MYB*, a proto-oncogene whose product is required for cell proliferation. Whether this particular method could provide long-term benefits in man remains to be seen, but the experiment does reinforce the view that anti-sense oligonucleotides may be highly versatile drugs.

7.2 The prospects for human gene therapy

The most ambitious and technically demanding application of genetic engineering techniques is the correction of human diseases by somatic cell gene therapy. The aim of such therapy is to insert an intact copy of the defective gene into somatic cells of the affected organism, and have it function stably such that the cognate protein will be synthesized and will correct the disease. It is important to distinguish this from germ-line therapy, where the aim is to correct the defect in the germ-line of the affected individual. There are major ethical problems with this latter concept.

One important long-term aim is to replace a mutated gene with its wild-type counterpart within a living cell. If such gene replacement could be achieved, problems of obtaining accurate expression of the newly introduced gene would be eliminated and there would be no risk of damage to the host genome caused by random integration of foreign DNA. The success in achieving homologous recombination in mouse cells in culture should act as a great spur to efforts to correct defective human genes by gene replacement. Most experiments so far performed in the mouse have been aimed at disrupting the homologous gene but it has also proven possible to achieve a form of gene replacement [7] using an allele exchange approach (see *Figure 6.1*).

Somatic therapy is of such enormous potential benefit that it need not await the achievement of true gene replacement in human cells. An alternative is gene supplementation, in which the transformed gene integrates randomly and produces its protein product. Gene supplementation seems achievable for some genetic diseases. The number of diseases where it may be contemplated is at present largely limited by the nature of the affected cell populations and whether the protein acts systemically. Gene supplementation should be applicable to diseases where the problem is caused by simple absence of the gene product, rather than by some specific malfunction of the mutated product, and where the precise level of expression of the introduced gene is not crucial. Suitable candidates include many recessive diseases.

The disease in which most progress has been made in establishing gene therapy is adenosine deaminase deficiency (ADA). This causes severe immunodeficiency and patients must remain isolated because of their high risk of contracting infections. In the attempts made thus far, the patient's lymphocytes were infected with retroviruses carrying the ADA gene under the control of a retroviral promoter and periodically re-injected into their bloodstream [8]. It is not yet known whether the *in vivo* supply of recombinant ADA protein will satisfy the patient's requirements and whether the therapy will have deleterious effects, but there is every cause for optimism that this will be an effective treatment. Fibroblasts are also relatively easy to isolate and re-implant, but thus far attempts to express secretory proteins, such as blood cell clotting factor IX, have resulted in detectable amounts of the product persisting for only a few days [9]. This could be due to the limited number of fibroblast precursor cells in the sample, or could reflect our relatively limited knowledge of the mechanisms controlling gene expression in these cells.

There are other diseases, where the protein product is required intracellularly, that also give cause for hope. This is particularly true for hematological disorders such as sickle cell anemia and the thalassemias. Bone marrow contains pluripotent stem cells, capable of generating every hematopoietic lineage, and there is considerable experience in the removal and re-implantation of bone marrow. Also, a vast amount of effort has gone into understanding globin gene expression and the sequences that direct erythroid gene expression are now relatively well defined. The fact that the globin genes are clustered, and that sequences at the periphery of the locus are required for high level expression of any gene within the cluster, is a complication (see Section 1.8). However, it has proven possible [10], in the case of the β-globin gene to construct a cassette that contains the gene and the peripheral sequences. This construct directs high level, erythroid-specific expression in transgenic mice. The major unresolved problem, before constructs such as this can be used in gene therapy, is that of introducing the gene into sufficient stem cells to repopulate the bone marrow. Here recent progress with retroviral vectors (Section 6.2.2) may hold the best hope of success. However, the requirement for precisely co-ordinated levels of expression of the α and β-globin genes, which are located on separate chromosomes, suggests that gene replacement rather than gene supplementation will be the best treatment for all the hemoglobinopathies.

Amongst the other target cells that are being investigated are muscle and liver cells, but here efforts are at a less advanced stage. If it were possible to use muscle cells in gene therapy then it might prove feasible to correct important diseases such as Duchenne muscular dystrophy (DMD). The DMD gene has been cloned; it encodes a cytoskeletal protein termed dystrophin. Myoblasts can readily be cultured *in vitro* and, when re-implanted in the host's tissue, they participate in the formation of myotubes. It is possible to infect myoblasts with retroviral vectors at a relatively high efficiency, but thus far dystrophin expression has been only transient [11].

In some cases attempts are being made to circumvent the biological problems of cell culture and re-implantation using novel delivery systems. Liver cells are, in theory, prime candidates for gene therapy because a large number of genetic diseases are caused by a deficiency in one or another liver enzyme. However, liver cells are extremely difficult to culture and, as an alternative to selecting a clone of transformed cells *in vitro*, a method of direct delivery of DNA into cells isolated by partial hepatectomy is under development [4]. It involves complexing the gene to asialoglycoproteins and it has proven possible to obtain transient expression using this method. In patients suffering from cystic fibrosis (CF), the lung cells secrete excessively sticky mucus which readily becomes infected and causes extreme difficulties in breathing. Here it is difficult to envisage any approach based on cell isolation and reimplantation. In rat experiments, an adenovirus vector has been used to transfect the CF gene into tracheal epithelium cells, and *in vitro* the CF gene was expressed. This raises the possibility of spraying such a recombinant virus into the lungs of affected individuals [12]. This approach has a significant in-built advantage over other methods of gene therapy, in that adenovirus does not integrate into the genome so that there is no risk of introducing an unwanted mutation by insertional gene disruption.

7.3 Genetic engineering in animals and plants

This book is primarily concerned with actual or potential applications of genetic engineering in medicine but in this final section we will briefly review some of the achievements of genetic engineering in non-human organisms. This has two purposes. First, some of the possibilities demonstrated in plants and animals may turn out to have a place in dealing with particular human problems. Secondly, genetically engineered non-human organisms are central to further medical progess as systems for producing new drugs and other medical products. *Table 7.1* summarizes some of the ways in which genes can be manipulated. In principle, any of these manipulations might be applied to any gene in any organism.

There are major ethical problems which preclude germ-line manipulation in humans, but the same constraints do not apply to plants and animals. The same techniques that are used to create transgenic animals have also now been applied to commercially important animals such as cows and sheep. This has a number of potential uses. It will be possible to change the genetic composition of these animals in a rational way, rather than relying on the selective breeding techniques that were used to create them. A very good example of the kind of changes that can be made is provided by an experiment in which the rat growth hormone gene was introduced into mice on a metallothionein promoter [13]. This yielded mice which grew, in some cases, to almost twice the normal size.

Another, more medically relevant, use of this technology will be to produce protein pharmaceuticals such as human growth hormone in the milk of cows or

Table 7.1:   *Ways in which a gene can be changed by genetic engineering*

Modifying gene expression:

Abolishing expression of a gene
Quantitatively reducing expression of a gene
Over-expressing a gene
Putting gene expression under the control of an external trigger
Changing the tissue specificity of expression
Changing the timing of expression during the cell cycle, during development, or in response
  to a stimulus

Modifying a gene product:

Giving it a new biological specificity
Abolishing one among several biological effects
Changing the rate at which it is metabolized or degraded
Changing the circumstances in which it is metabolized or degraded
Inducing resistance or sensitivity to a drug
Changing its intracellular location
Causing or preventing secretion of the product

---

sheep. The genes that encode the most abundant proteins in milk, the caseins, have been cloned. When the casein promoter is coupled to the coding region of a gene it produces large amounts of protein at a very low relative price. Crop plants can also be stably transformed and they offer another potential route to preparing pharmacologically important proteins in large amounts. Plants would have one advantage in that animal proteins which might produce unwanted effects in the mammary gland of a cow would be very unlikely to do so in a plant cell. They would, however, suffer from the disadvantage that the proteins would not be glycosylated in the same way as in an animal cell.

The wider and much greater promise of plant molecular biology is to produce new plant strains that are better fitted to their environment, that are disease resistant or that produce a more nutritious food product. If this can be achieved then the impact on health in the developing world will be enormous.

In principle the possibilities for genetic engineering are almost limitless. Organisms might be made with any mix and match of desired characteristics. Even conventional breeding programs within an existing species can realize a vast reservoir of potentially useful variation – think of all the different breeds of dogs, whose varying physiques and temperaments were all potentially present in the original wild stock. Individual genes or their products can be modified in any of the ways summarized in *Table 7.1*. Differences between species may have a surprisingly small genetic basis. Humans and mice share the overwhelming majority of all their genes, usually in conserved linkage groups. The differences between humans and mice, which seem so great to us, probably depend mainly on quite small differences in the relative timing of developmental programs which use the same genes. Such timing is potentially amenable to manipulation. Beyond this, new genes can be invented by combining functional modules of existing genes, mimicking the evolutionary exon shuffling which is believed to have created the diversity of existing genes. If this is still insufficient, we can imagine creating entirely novel genes.

Thus it seems likely that genetic engineering has the potential to produce organisms to any desired specification. What use society chooses to make of this power is, of course, not a matter for the scientists alone. Like any new technology, genetic engineering has potential for good and evil. However, it must be said that, compared to many other technologies, with genetic engineering the benefits are easier to describe and realize than most of the abuses.

References

1. Weinberg, R.A. (1989) *Cancer Res.*, **49**, 3713.
2. Sukumar, A. (1989) *Current Topics Microbiol. Immunol.*, **148**, 93.
3. Vogelstein, B. (1990) *Nature*, **348**, 681.
4. Friedman, T. (1989) *Science*, **244**, 1275.
5. Eck, S.L. and Nabel, G.J. (1991) *Current Opinion in Biotech.*, **2**, 897.
6. Simons, M., Edelman, E.R., DeKeyser, J.L., Langer, R. and Rosenberg, R.D. (1992) *Nature*, **359**, 67.
7. Hasty, P., Ramirez-Solis, R., Krumlauff, R. and Bradley, A. (1991) *Nature*, **350**, 243.
8. Marwick, C. (1991) J. Am. Med. Assoc., **256**, 2311.
9. Palmer, T.D. (1991) Proc. Natl Acad. Sci. USA, **88**, 1330.
10. Grosveld, F., Assendfelt, G.B., Greaves, D.R. and Kollias, G. (1987) *Cell*, **51**, 975.
11. Miranda, A.F. (1990) Adv. Exp. Med. Biol., **280**, 205.

12. Rosenfeld, M.A., Siegfried, W., Yoshimura, K., *et al.* (1991) *Science*, **252**, 431.
13. Palmiter, R.D., Brinster, R.L., Hammer, R.E., Trumbauer, M., Rosenfeld, M.G., Birnberg, N.C. and Evans, R.M. (1982) *Nature*, **300**, 611.

Further reading

Capecchi M. (1989) Altering the Genome by Homologous Recombination. *Science*, **244**, 1288–1292.

Cossman, J. and Schlegel, R. (1991) p53 in the diagnosis of human neoplasia. *J. Natl. Cancer Inst.*, **83**, 980–981.

(1991) *Current Opinions in Genetic and Development*, **1**, n.1.

Friedman, T. (1989) Progress towards human gene therapy. *Science*, **244**, 1275–1281.

Levine, F. and Friedman, T. (1991) Gene therapy techniques. *Current Opinion in Biotech.*, **2**, 840–844.

Miller, A.D. (1992) Human gene therapy comes of age, *Nature*, **357**, 455.

Ponder, B.A.J. (1990) Inherited predisposition to cancer. *BAJ* **6**, 213–218.

Verma I.M. (1990) Gene therapy, *Sci. Am.*, **262**, 68–84

Westphal, H. (1991) Mouse models of human diseases. *Current Opinion in Biotech.*, **2**, 830–833.

APPENDIX A. GLOSSARY

Activation domain: the part of a transcription factor which is believed to recruit RNA polymerase II to the 5' end of the gene.

Allele exchange: a method of gene replacement which can be used to introduce mutations into the genome of an organism.

Annealing or hybridization: the process whereby two initially separate, but complementary, nucleic acid molecules form a double-stranded structure.

Anti-sense RNA: this term normally refers to the RNA copy of a specific mRNA. Anti-sense RNA is used *in vitro* as a probe for hybridization and *in vivo* as a way of blocking gene transcription.

Autonomously replicating sequences (ARSs): segments of DNA that, when inserted into a circular plasmid, confer the ability to replicate extra-chromosomally in yeast.

Background: a problem in most cloning procedures, where some (or in worst cases all) of the bacterial colonies which grow up do not contain vector molecules.

Cap: a methylated guanine residue which is added to the 5' end of eukaryotic mRNAs in a post-transcriptional reaction.

Cap site: the start site of transcription of a eukaryotic gene.

cDNA clone: a recombinant molecule containing the double-stranded DNA copy of a mRNA sequence.

Chimera: an embryo produced by introducing pluripotential cells into a normal embryo.

Chromosome walking: sequential cloning steps designed to bridge the gap between a probe that is to hand and a linked, but not immediately adjacent, region of a chromosome.

Codon: a triplet of nucleotides in a mRNA molecule that specifies the insertion of a particular amino acid into the growing polypeptide chain.

Cosmid: a plasmid vector that contains *cos* sites and which can therefore be packaged into pseudo-viral particles.

Cos **sites:** the mutually adhesive termini of bacteriophage lambda.

Co-transformation: a procedure whereby two different DNA molecules, only one of which need contain a selectable marker, are mixed together and introduced into the genome of eukaryotic cells.

CpG islands: a region of genomic DNA that contains a very high proportion of CpG residues, a di-nucleotide that is generally highly under-represented in mammalian DNA. CpG islands are often located at the 5' ends of genes.

Cross-hybridization: the annealing of two nucleic acid molecules which are not perfectly complementary.

Deletion series: a set of clones, all derived from the same initial recombinant, but in which the insert lacks sequences at one of its ends because of treatment with an exonuclease.

Denaturation: a process whereby the two strands of a double-stranded nucleic acid molecule come apart as a result of heating or exposure to alkali conditions.

Differential screening: a method of screening cDNA libraries whereby two probes, differing in one or just a few sequences, are used as probes in two parallel *in-situ* hybridizations using duplicate lifts of the library.

Differential splicing: a form of splicing wherein the combination of introns which are removed varies between cell types, allowing a single gene to produce transcripts which differ in primary structure and therefore in coding potential.

DNA binding domain: the part of a transcription factor which tethers the protein to its cognate regulatory element.

DNA polymerase: an enzyme which copies a DNA or an RNA molecule to produce a DNA copy.

Ectopic gene expression: the expression of a gene in a tissue where it is normally inactive. This is normally achieved by linking the coding region of the gene to a heterologous promoter.

Efficiency elements: eukaryotic gene regulatory elements that act to increase the overall rate of gene transcription but do not directly confer tissue specificity of expression.

Electroporation: a method which uses an electrical pulse to introduce DNA into cells for transformation.

Endonuclease: a nuclease which cuts a nucleic acid molecule by cleaving between two internal residues.

Enhancers: eukaryotic gene regulatory elements that can confer tissue specificity of expression and which can activate gene transcription even when situated great distances away from the cap site of the gene.

Exon: a part of a nuclear RNA precursor which is joined together with other exons within the same RNA by splicing. The product of this reaction, the mRNA, is then exported to the cytoplasm.

Exonuclease: a DNA exonuclease which degrades a double-stranded DNA molecule by progressively removing nucleotides from its two ends.

Expression screening: a method of screening for a specific cDNA clone where the cDNA is inserted next to a promoter active in the host cell and an immunoassay or bioassay is used to detect the required clone.

Fluorescent *in-situ* hybridization (FISH): a technique whereby a probe that can be detected using a fluorescently labeled reagent is annealed to chromosomes with the aim of localizing the target sequences.

Gene disruption: a method of inactivating a gene, within a living cell, by transforming it with a construct which is able to undergo homologous recombination.

Gene fusion: a DNA segment containing parts of different genes, e.g. the promoter of one gene and the coding region of another gene.

Genome: the sum of the genetic information necessary to specify the formation of a living organism or a virus.

Genomic clone: a recombinant molecule containing genomic DNA, normally the term refers to a clone containing a gene and a variable amount of flanking DNA depending upon the method of cloning used.

Germ-line chimera or germ-line mosaic: a transgenic mouse in which some or all of the germ cells contain the transgene and which is therefore able to pass the gene on to its progeny.

Histones: basic proteins which associate with, and package, eukaryotic DNA as part of the formation of chromosomal DNA.

Homeobox (*Hox*) domain: a DNA binding domain first discovered in *Drosophila* homeotic genes.

Homeotic genes: genes which specify segment identity in *Drosophila melanogaster*.

Insert: a target for isolation by DNA cloning, e.g. the cDNA copy of an mRNA.

Intron: part of a nuclear RNA precursor which is removed and degraded within the nucleus to yield mRNA.

In-vitro packaging: a method whereby DNA flanked by phage packaging signals is encapsidated *in vitro*.

Lift: a replica copy of a bacteriological plate bearing bacterial colonies or phage plaques made by touching an inert filter on to the plates surface.

Linkage analysis: the use of polymorphic variation to estimate genetic distance.

Locus control regions (LCRs) or domain control regions (DCRs): regions of the genome necessary for activation of the entire β-globin gene locus in erythroid cells. Other gene clusters almost certainly possess LCRs but this has yet to be demonstrated.

Lod score: the statistical probability that two markers which appear to segregate together are actually linked.

Maxam and Gilbert (chemical degradation) procedure: a DNA sequence analysis method.

MCS (multi-cloning site or polylinker): a short DNA sequence, found in most vectors in common use, which contains many closely spaced restriction enzyme cleavage sites.

Melting temperature or T_m: the temperature at which the two strands of a double-stranded nucleic acid molecule come apart.

Messenger RNA (mRNA): an RNA molecule that contains the genetic information necessary to encode a protein.

Microsatellites: a class of highly variable markers in human DNA which are composed of di-, tri- and tetra-nucleotide repeats, for example $(CA)_n$ or $(CCA)_n$.

Mini-preparations: DNA prepared from small bacterial cultures derived from individual colonies in a cloning experiment.

Mis-matches: non-complementary base pairs in a nucelic acid duplex.

Mosaic: a transgenic animal in which the transgene is present in only a fraction of cells within the animal.

Multi-allelic variation: genes or DNA segments showing high levels of polymorphic variation exemplified by VNTRs and microsatellites.

5′ and 3′ non-coding regions: untranslated regions at the 5′ and 3′ ends of mRNA sequences.

Northern blotting: a process whereby RNA molecules are separated by gel electrophoresis, transferred to a filter and hybridized with a specific probe with the aim of detecting complementary target molecules.

Nucleosome: an octamer, composed of two molecules of each of the core histones, H2A, H2B, H3 and H4, forming a disc-shaped structure with approximately 140 bp of DNA wrapped around it.

Open reading frames (ORFs): a region of DNA sequence that does not contain one of the three stop codons and therefore has the potential to encode a protein.

Operons: closely spaced bacterial genes which function in a common metabolic pathway.

Plaque: an area of a bacteriological plate where the bacteria are dead, or grow slowly because of infection by a virus.

Plasmid: extrachromosomal, circular DNA molecules found in prokaryotes and some lower eukaryotes.

Pluripotency: this is the defining property of a cell, e.g. a mouse ES cell, that is able to participate in embryonic development.

Poly(A) tail: a tract of A residues of approximately 100–200 nucleotides in length that is added enzymatically to the 3′ end of a mRNA.

Polymerase chain reaction (PCR): a method of specifically copying, and amplifying, a part of a nucleic acid chain.

Polysome: a translationally active complex comprised of a mRNA and its associated ribosomes.

Pooled (amplified) libraries: for long-term storage phage particles or bacterial colonies on bacteriological plates are eluted in a mixed pool, containing representatives of every different recombinant on the plate.

Primer: a short nucleic acid molecule which, when annealed to a complementary template strand, provides a 3′ terminus suitable for copying by a DNA polymerase.

Primer extension: a method of establishing the start site of transcription of a gene.

Probe: a nucleic acid sequence that is complementary to part or all of a target which is to be detected by hybridization. It is usually 'tagged' by the incorporation either of radioactively labeled nucleotides or of nucleotides which are chemically modified in such a way that they can be identified immunologically.

Pulse field gel electrophoresis (PFGE): a form of gel electrophoresis that allows extremely long DNA molecules to be separated from one another.

Rare cutter: a restriction enzyme which recognises sites that are infrequently represented.

Replication origin: a segment of DNA that acts as the start site of DNA replication.

Restriction enzymes: enzymes which cleave double-stranded DNA into discrete pieces by cleaving at defined recognition sequences.

Restriction fragment-length polymorphism (RFLP): a localized difference in the genome structure of individuals within a population that produces a difference in their restriction maps.

Restriction map: a physical map of a piece of DNA showing the position of cleavage of one or more restriction enzymes.

Ribosomal RNA (rRNA): an RNA molecule that forms part of the ribosome, the bipartite structure where mRNAs are translated into proteins.

RNA polymerase: an enzyme which copies a DNA or an RNA molecule to produce its RNA copy.

Sanger (dideoxy or chain termination) procedure: the standard method of DNA sequence analysis.

Shuttle vector: a vector that is able to transform both *E. coli* and some other organism; constructions and mutations can be made in *E. coli* and the effects studied in the alternate host.

Signal peptide: a stretch of predominantly hydrophobic amino acids which directs the polysomes to the endoplasmic reticulum during translation.

SI mapping: a method of determining the transcriptional organization of a gene.

Southern blotting: a process whereby DNA molecules are separated by gel electrophoresis, transferred to a filter and hybridized with a specific probe with the aim of detecting complementary target molecules.

Specific activity: the amount of radioactivity incorporated per weight of probe DNA present in a labeling reaction.

Splicing: the series of steps within the nucleus whereby introns are removed from the nuclear precursor to form a mRNA and the exons are joined together.

Stringency of hybridization: the combination of temperature, salt and formamide concentration which determine the degree of cross-hybridization that can occur in an annealing reaction or during post-hybridization washing.

Subcloning: the process wherein a purified DNA molecule is inserted into a vector and isolated by gene cloning.

Subtraction hybridization (subtraction enrichment): a method of screening cDNA libraries whereby two probes, differing in one or just a few sequences, are hybridized together so that the sequences common to the two populations anneal. The non-annealed sequences are then used to prepare a cDNA library and/or a probe for screening.

Tail blot: a procedure for analyzing transgenic mice for the presence of a transgene.

Target: a nucleic acid molecule which is to be detected in a hybridization reaction or which is to be isolated by molecular cloning.

Transcription: the process whereby a DNA molecule is copied into RNA.

Transcription factor: a protein that regulates or facilitates the transcription of a gene.

Transfer RNA (tRNA): the intermediaries which carry amino acids to the ribosome and which direct their position of insertion into the polypeptide chain.

Transformation (DNA transformation): the process whereby a DNA molecule is introduced into a living cell. The term is normally reserved for those cases where the DNA molecule has the capacity to be stably maintained in the host cell.

Transgenic mice: mice containing an experimentally introduced DNA segment (a transgene) integrated into the genome of some or all of their cells.

Transient expression: a technique in which DNA is introduced into eukaryotic cells and its transcription is analyzed after it has reached the nucleus, but before it has integrated into the genome.

Translation: the process whereby rRNAs, and their associated proteins, decode the linear sequence of information contained within the mRNA to form a protein.

Unequal crossing over: a non-reciprocal exchange of genetic material at meiosis.

Variable number tandem repeats (VNTRs): polymorphic markers composed of multiple tandem repeats of a short DNA segment.

Vector: a DNA molecule with the ability to replicate in its cognate host cell and which contains a genetic marker that allows for selection of cells that contain it.

Yeast artificial chromosomes (YACs): recombinant DNA molecules that contain very large DNA inserts and which replicate in yeast cells as linear, mini-chromosomes.

INDEX